International Nuclear Science, Technology and Engineering Conference (iNuSTEC 2021)

AA007322

IOP Conference Series: Materials Science and Engineering
Volume 1231

Online
9 – 12 October 2021

Editors:

Nahrul Khair Alang Md Rashid
Khaidzir Hamzah
Mohsin Mohd Sies
Jasman Zainal
Muhammad Arif Sazali
Muhammad Syahir Sarkawi
Nur Syazwani Mohd Ali
Khairulnadzmi Jamaluddin
Abdul Aziz Mohamed
Faridah Mohamad Idris
Julie Andrianny Murshidi
Faizal K P Kunchi Mohamed
Hassan Mohamed
Abu Hassan Husin

ISBN: 978-1-7138-4689-5
ISSN: 1757-8981

Printed from e-media with permission by:

Curran Associates, Inc.
57 Morehouse Lane
Red Hook, NY 12571

Some format issues inherent in the e-media version may also appear in this print version.

Institute of Physics
Dirac House, Temple Back
Bristol BS1 6BE UK

Phone: 44 1 17 929 7481
Fax: 44 1 17 920 0979

techtracking@iop.org

Additional copies of this publication are available from:

Curran Associates, Inc.
57 Morehouse Lane
Red Hook, NY 12571 USA
Phone: 845-758-0400
Fax: 845-758-2633
Email: curran@proceedings.com
Web: www.proceedings.com

TABLE OF CONTENTS

IOP Conference Series: Materials Science and Engineering

International Nuclear Science, Technology and Engineering Conference 2021 (iNuSTEC2021)

"Nuclear Science and Technology for Socio-economic Development"

Universiti Teknologi Malaysia (UTM)
Skudai, Johor, Malaysia
10-12 October 2021

Editors

Nahrul Khair Alang Md Rashid, Universiti Teknologi Malaysia
Khaidzir Hamzah, Universiti Teknologi Malaysia
Mohsin Mohd Sies, Universiti Teknologi Malaysia
Jasman Zainal, Universiti Teknologi Malaysia
Muhammad Arif Sazali, Universiti Teknologi Malaysia
Muhammad Syahir Sarkawi, Universiti Teknologi Malaysia
Nur Syazwani Mohd Ali, Universiti Teknologi Malaysia
Khairulnadzmi Jamaluddin, Universiti Teknologi Malaysia
Abdul Aziz Mohamed, Malaysian Nuclear Society
Faridah Mohamad Idris, Malaysian Nuclear Agency
Julie Andrianny Murshidi, Malaysian Nuclear Agency
Faizal K P Kunchi Mohamed, Universiti Kebangsaan Malaysia
Hassan Mohamed, Universiti Tenaga Nasional
Abu Hassan Husin, Universiti Teknologi MARA

All papers have been peer-reviewed

Supporting/Sponsoring Organisations

Ministry of Education Malaysia; Ministry of Science, Technology and Innovation Malaysia; Ministry of Energy and Natural Resources; Malaysian Nuclear Agency; ASME Malaysia Section; Malaysian Welding and Joining Society; Reactor Interest Group; Asia-Oceania Neutron Scattering Association, Japan Atomic Energy Society; Advanced Nuclear Engineering Research Group

PREFACE

It is my pleasure to welcome all the distinguished guests, keynote speakers, and participants. Universiti Teknologi Malaysia (UTM) is truly honored to host the three-day virtual conference of the International Nuclear Science, Technology and Engineering Conference 2021 (iNuSTEC2021). The iNuSTEC2021 is a conference organized by the Malaysian Nuclear Society (MNS) through collaborations with local universities as a platform for scientists, researchers, and interested parties around the world to share experiences and exchange their knowledge as well as new findings on nuclear, quantum, and radiation-related technologies for the growth of nuclear science, technology, and engineering.

I wish a special welcome to all the eminent speakers from Malaysia and nuclear experts from China, South Korea, Japan, Austria, and Pakistan to iNuSTEC2021, and thus making this conference a truly international conference in spirit. I'm very glad and thankful that you are willing to spend your precious time to be with us and share your knowledge and vast experiences with us, especially for the young budding nuclear engineers in Malaysia.

I would like to take this opportunity to share the Universiti Teknologi Malaysia's involvement in nuclear education. UTM is currently the only university in Malaysia that offers bachelor's degree in Nuclear Engineering starting in 2012. In 2021, we have more than 100 alumni and 60 current students. Even though we do not have nuclear power, our alumni were able to secure employment in other engineering fields. Some were able to secure international scholarships to pursue their postgraduate study in universities such as the Tsinghua University of China, Korea Advanced Institute of Science and Technology (KAIST), and Kepco International Nuclear Graduate School (KINGS) of South Korea, proving the excellent quality in our local students that we have nurtured. The undergraduate students are also active in their Nuclear Engineering Students Society. They have collaborated with local and international experts in conducting their events to disseminate knowledge on nuclear technology to the public. I was also informed that they were responsible for conducting the conference concurrent events. Congratulations to all who were involved. Additionally, a Memorandum of Understanding (MOU) between UTM and Nuclear Malaysia was formalized, and it has been renewed recently. This is done to strengthen the cooperation and collaboration with Nuclear Malaysia since there are no nuclear facilities at UTM, providing the students with the opportunity to use their facilities. Exposing students to the TRIGA Mark II research reactor and NDT laboratories help our students to understand the physics and applications of nuclear technologies better. Besides, research activities are also conducted with Nuclear Malaysia facilities for the benefit of staff and students. Hopefully, these students will be the ambassadors in supporting the establishment of nuclear power in Malaysia in the future.

Congratulations and kudos to the Organizing Committee for hosting this meaningful event concurrently with Debat Nuklear Antara Varsiti 2021 and Nuclear Youth Competition 2021. My deepest appreciations to keynote speakers, technical co-sponsors, and all supporters for their contributions to making sure the event is successful. Last but not least, I would like to thank the presenters and participants for your participation, and I believe your contribution will support the advancement of knowledge in nuclear technology.

Thank you.

Professor Datuk Ts Dr Ahmad Fauzi Ismail
Vice-Chancellor
Universiti Teknologi Malaysia
Patron of iNuSTEC2021

IOP Conf. Series: Materials Science and Engineering 1231 (2022) 011001 doi:10.1088/1757-899X/1231/1/011001

MESSAGE FROM:
PRESIDENT, MALAYSIAN NUCLEAR SOCIETY (MNS)

I am really pleased to welcome all keynote speakers, sponsors, presenters, and participants to the International Nuclear Science, Technology and Engineering Conference (iNuSTEC2021). Ideally, the iNuSTEC2021 is organized to enhance the importance of nuclear science, technology, and engineering in developing socio-economic in Malaysia. Malaysian Nuclear Society (MNS) is truly honored to organize this event for the third time with Universiti Teknologi Malaysia (UTM). Successful conferences from previous years, at both national and international levels, have brought this event to be more meaningful to strengthen nuclear development for global growth.

With the support from the American Society of Mechanical Engineers (ASME) Malaysia Section, and Malaysian Welding and Joining Society (MWJS), the MNS is indeed honored to co-organize this event with UTM and with organizations such as Nuclear Malaysia Agency, UiTM, UKM, UNITEN, UM, USIM and UTHM. The affiliation of the Nuclear Engineering Students Society (NESS) gives inception to a new MNS Chapter and has moved towards greater strength in nuclear fields. In the future, MNS members shall strive to make the MNS objectives and visions better and stronger.

This conference will be participated by experienced and young researchers that can give a platform for them to discuss and debate on the latest development in all areas of nuclear technology applications for friendly purposes at the international level. This would spur and catalyse further research and development leading to the sustainability and improvement of the quality of life without compromising the quality of the environment in Malaysia and the world.

MNS is a unique non-governmental organization that is formed and headed by individuals that are aligned to the beneficial usage of nuclear science, engineering, and technology which has been incorporated under the Malaysian Society Act. The MNS has been actively promoting the information, advancement, popularization, and understanding of nuclear the public in Malaysia and this conference is one of the ways of realizing such aim. This conference is also served as a public acceptance program to introduce the benefits and advantages of the use of nuclear science, engineering, and technology to the public on a large scale. On top of that MNS also published twice a year on regular basis the Journal of Nuclear and Related Technologies (JNRT) and Buletin Nuklear Malaysia (BNM) for researchers to share their research findings with the community. We also publish iNuSTEC proceedings on yearly basis.

I wish to express my heartful thanks to all speakers for your willingness to share your experience with the participants and presenters for the development of the nuclear field. The contribution is not only coming from the Malaysian scientific community but also distinguished guest speakers of iNuSTEC2021. I also hope my appreciation and thanks to all sponsors for their generous supports. May all of the efforts bring benefits to all of us, and I wish everyone a successful conference.

Thank you.

Associate Professor Dr Abdul Aziz Mohamed
President
Malaysian Nuclear Society

IOP Conf. Series: Materials Science and Engineering 1231 (2022) 011001 doi:10.1088/1757-899X/1231/1/011001

CONTENTS

PAPERS

PROCEEDING EDITORIAL COMMITTEE

Nahrul Khair Alang Md Rashid (*UTM*)
Muhammad Arif Sazali (*UTM*)
Muhammad Syahir Sarkawi (*UTM*)
Nur Syazwani Mohd Ali (*UTM*)
Nor Afifah Basri (*UTM*)
Khairulnadzmi Jamaluddin (*UTM*)
Abdul Aziz Mohamed (*MNS*)
Faridah Mohamad Idris (*MNS/Nuklear Malaysia*)
Nasri A. Hamid (*MNS/UNITEN*)
Mohd Syukri Yahya (*MNS/UNITEN*)
Hassan Mohamed (*MNS/UNITEN*)
Julie Andrianny Murshidi (*MNS/Nuklear Malaysia*)
Juniza Md Saad (*MNS/UPM*)
Mohd Idzat Idris (*MNS/UKM*)
Imran Yusuff (*MNS/UKM*)
Abu Hassan Husin (*UiTM*)
Khaidzir Hamzah (*UTM*)

IOP Conf. Series: Materials Science and Engineering 1231 (2022) 011001 doi:10.1088/1757-899X/1231/1/011001

iNuSTEC2021 ORGANISING COMMITTEE

Patron
Professor Datuk Ts Dr Ahmad Fauzi Ismail
Vice Chancellor, Universiti Teknologi Malaysia (UTM)

Chairman
Professor Ir Dr Nahrul Khair Alang Md Rashid
Nuclear Engineering Programme, SCEE, FE, UTM

Co-Chairman
Associate Prof Dr Aziz Mohamed
President, Malaysian Nuclear Society (MNS)

Deputy Chairman
Associate Prof Dr Khaidzir Hamzah
Nuclear Engineering Programme, SCEE, FE, UTM

Secretary
Dr Nur Syazwani Mohd Ali
Nuclear Engineering Programme, SCEE, FE, UTM
Dr Faridah Mohamad Idris
Secretary, MNS

Treasurer
Dr Julie Andrianny Murshidi
Treasurer, MNS

IOP Conf. Series: Materials Science and Engineering 1231 (2022) 011001 doi:10.1088/1757-899X/1231/1/011001

INTERNATIONAL ADVISORY COMMITTEE

Professor Emiretus Dato' Dr Noramly Muslim, *MNS Honorary Fellow*
Dr Robert Robinson, *Asia-Oceania Neutron Scattering Association (AONSA)*
Dr Kiyonobu Yamashita, *JAIF International Cooperation Center, JAPAN*
Professor Dr Lee Deokjung, *Ulsan National Institute of Science and Technology, SOUTH KOREA*
Professor Dr Kim Hee Reyoung, *Ulsan National Institute of Science and Technology, SOUTH KOREA*
Mr Ding Jian, *CNNC China Zhongyuan Engineering Corporation, CHINA*
Dr Andreas Keilbach, *Anton Paar GmbH, AUSTRIA*
Professor Dr Zafar Koreshi, *Air University, PAKISTAN*

NATIONAL ADVISORY COMMITTEE

Ts Dr Siti Aia'sah Hashim, *Nuklear Malaysia*
Professor Ts Dr Ruzairi Abdul Rahim, *Universiti Teknologi Malaysia*
Professor Dr Mohd Ghazali Mohd Nawawi, *Universiti Teknologi Malaysia*
Dr Hasrinah Hasbullah, *Universiti Teknologi Malaysia*
Dr Mohsin Mohd Sies, *Universiti Teknologi Malaysia*
Associate Professor Dr Abu Hassan Husin, *Universiti Teknologi MARA*
Ts Saharuddin Kambari, *Malaysian Welding and Joining Society (MWJS)*
Professor Dr Faizal K P Kunchi Mohamed, *Universiti Kebangsaan Malaysia*

TECHNICAL COMMITTEE

Technical Unit
Dr Muhammad Syahir Sarkawi, *Universiti Teknologi Malaysia*
Mohd Shafiq Md Hidiah, *Universiti Teknologi Malaysia*
Abdul Khalil Abdollah, *Universiti Teknologi Malaysia*
Choo Kai Jie, *Nuclear Engineering Students Society UTM*
Chin Lok Sheng, *Nuclear Engineering Students Society UTM*
Tengku Nur Aishah Balqis Tengku Asmansany, *Nuclear Engineering Students Society UTM*

Program and Multimedia Unit
Dr Muhammad Arif Sazali, *Universiti Teknologi Malaysia*
Dr Khairulnadzmi Jamaluddin, *Universiti Teknologi Malaysia*
Chai Chan Ful, *Nuclear Engineering Students Society UTM*
Muhammad Amir Asyraf Mahadzer, *Nuclear Engineering Students Society UTM*
Fadhila Jamaluddin, *Nuclear Engineering Students Society UTM*

Supporting Unit
Dr Jasman Zainal, *Universiti Teknologi Malaysia*
Dr Wan Norhayati Wan Salleh, *Universiti Teknologi Malaysia*
Dr Zulhairun Abdul Karim, *Universiti Teknologi Malaysia*
Mohammed Khairul Anwar Sulaiman, *Universiti Teknologi Malaysia*
Marina Mukhtar Dahlan, *Universiti Teknologi Malaysia*
Azila Draman, *Universiti Teknologi Malaysia*
Sarimah Satam, *Universiti Teknologi Malaysia*
D. Dahyuna Mohd Yunos, *Universiti Teknologi Malaysia*
Ardiyana Mohamad Suhaimi, *Universiti Teknologi Malaysia*

DEBAT NUKLEAR ANTARA VARSITI 2021 SECRETARIAT

Dr Amira Othman, *Universiti Tun Hussein Onn Malaysia*
Muhammad Asyraf Muhamad Sidek, *Nuclear Engineering Students Society UTM*
Muhammad Izzat Aiman Zakaria, *Nuclear Engineering Students Society UTM*
Adam Haziq Mohamad Fahmi, *Nuclear Engineering Students Society UTM*
Nur Fatin Athirah Amran, *Nuclear Engineering Students Society UTM*
Ko Yu Yun, *Nuclear Engineering Students Society UTM*
Mahirah Hj Nasir, *Nuclear Engineering Students Society UTM*

IOP Conf. Series: Materials Science and Engineering 1231 (2022) 011001 doi:10.1088/1757-899X/1231/1/011001

NUCLEAR YOUTH COMPETITION 2021 SECRETARIAT

Dr Nor Afifah Basri, *Universiti Teknologi Malaysia*
Nur Afiah Syakirah Roslan, *Nuclear Engineering Students Society UTM*
Nisa Faizah Yakob, *Nuclear Engineering Students Society UTM*
Ainur Najwa Mohamad Pouzi, *Nuclear Engineering Students Society UTM*
Devi Durgaashini Ramesh, *Nuclear Engineering Students Society UTM*
Teh Zhi Hui, *Nuclear Engineering Students Society UTM*
Tan Lun Yong, *Nuclear Engineering Students Society UTM*
Bernard Choong Foo Tin, *Nuclear Engineering Students Society UTM*

LIVE STREAMING SECRETARIAT

Mohamad Hafhiz Abd Aziz, *Hafsha Global Resources Malaysia*

INUSTEC 2021 AWARDS AND RECOGNITION

Congratulations All Winners

DEBAT NUKLEAR ANTARA VARSITI 2021

Champion:
Universiti Tenaga Nasional
Muhammad Syazran Haikal Bin Azlan
Nur Aziean Binti Abdullah
Mohamad Luqman Hafiz Bin Md Hanif
Wan Qistina Binti Wan Izahan

1st Runner Up:
Universiti Kebangsaan Malaysia (UKM)
Maizatul Amira Binti Muhamad
Nur' Atikah Binti Zakaria
Anis Farhana Binti Mohd Rosoliza
Nur Amira Hidayah

Consolation Prize 1:
Universiti Sultan Zainal Abidin (UniSZA)
Muhammad Hafiqrul Zuhair Bin Husri
Mohamad Izzat Bin Zulkifli
Nur Athirah Binti Mat Arman
Nur Hanisah Binti Mohamad Hadi Panisevom

Consolation Prize 2:
Universiti Teknologi MARA (UiTM)
Muhammad Alif Naqiuddin Bin Mohd Adzha
Muhammad Harits Ikhwan Bin Khairul Azman
Fathin Kamilia Binti Mohamad Hussin
Nur Anis Anesha Binti Ahmad Naim

IOP Conf. Series: Materials Science and Engineering 1231 (2022) 011001 doi:10.1088/1757-899X/1231/1/011001

NUCLEAR YOUTH COMPETITION 2021

Champion:
Ainur Najwa Bt Mohd. Pouzi (*UTM*)
Devi Durgaashini A/P Ramesh (*UTM*)
Ko Yu Yun (*UTM*)
Chin Lok Sheng (*UTM*)

1st Runner Up:
Nur Afiah Syakirah Bt Roslan (*UTM*)
Gilbert Raphael Bong (*UKM*)
Muhammad 'Adli Bin Nor Azman (*UTM*)
Choo Kai Jie (*UTM*)

Consolation Prize 1:
Nisa Faizah Bt Yakob (*UTM*)
Nurliyana Athirah Binti Ibrahim (*UTM*)
Pang Chern Hong (*UTM*)

NUCLEAR QUIZ CHALLENGE

Champion: Pang Chern Hong (*UTM*)
1st Runner Up: Chan Chai Ful (*UTM*)
2nd Runner Up: Nisa Faizah Bt Yakob (*UTM*)

PERTANDINGAN POSTER/VIDEO

Chan Chai Ful (*UTM*)

ACKNOWLEDGEMENT

iNuSTEC2021 is jointly organized by Malaysian Nuclear Society (MNS)
and Universiti Teknologi Malaysia (UTM)

In Collaboration with

Malaysian Nuclear Agency (Nuklear Malaysia)
Universiti Tenaga Nasional (UNITEN)
Universiti Kebangsaan Malaysia (UKM)
Universiti Teknologi MARA (UiTM)
Universiti Islam Antarabangsa Malaysia (IIUM)
Universiti Malaya (UM)
Universiti Sains Islam Malaysia (USIM)
Universiti Tun Hussien Onn Malaysia (UTHM)
Arey Solutions Sdn Bhd

Co-organized by
Nuclear Engineering Students Society (NESS)
Institute of Nuclear Engineers (INE MNS)
Kelab Sains Nuklear (KESAN)
Malaysian Nuclear Professional Chapter (MNPC MNS)

Supported by
ASME Malaysia Section
Malaysian Welding and Joining Society

MNS COUNCIL MEMBERS 2021/2023

Advisor / Past President	Prof. Dr. Nahrul Khair Bin Alang Md Rashid (*UTM*)
President	Dr. Abdul Aziz Bin Mohamed (*UNITEN*)
Vice President	Assoc. Prof. Dr. Faizal Bin K.P. Kunchi Mohammed (*UKM*)
Secretary	Dr. Faridah Bt Mohamad Idris (*Nuklear Malaysia*)
Deputy Secretary	Anas Bin Mohamad Fauzi (*UNITEN*)
Treasurer	Dr. Julie Andrianny Bt. Murshidi (*Nuklear Malaysia*)
Deputy Treasurer	Khair'ah Binti Yazid@Khalid (*Nuklear Malaysia*)

Members
Ts. Dr. Hassan Bin Mohamed (*UNITEN*)
Dr. Julia Bt. Abdul Karim (*Nuklear Malaysia*)
Dr. Shukri Mohd (*Nuklear Malaysia*)
Dr. Muhammad Rawi B. Mohamed Zin (*Nuklear Malaysia*)
Dr. Mohd Syukri Yahya (*UNITEN*)

Auditors
Khairul Azhar Bin Abdul Halim (*Nuklear Malaysia*)
Ikmal Azrai Bin Abdul Aziz (*Arey Solutions*)

MNS Chapters
RadTech (*Radiation Tech*)
XApp (*X-ray App*)
NuEnergy (*Nuclear Energy*)
Neutron Photon (*Neutron- Photon*)
Student Chapter (*UKM, UiTM, UNITEN, UTM - NESS*)
Women in Nuclear, Malaysia (*WiN Malaysia*)
Institute of Nuclear Engineers (*INE MNS*)
Malaysian Nuclear Professional Chapter (*MNPC MNS*)

MNPC-MNS COMMITTEE MEMBERS 2021/2023

Advisor / President MNS
Dr. Abdul Aziz Bin Mohamed

President
Ts. Razali Bin Harun (*REM*)

V. President
Khairul Izzat Bin Abdul Aziz (*Reltech Lab*)

Secretary
Mohd Taha Bin Hj. Suja' (*REM*)

Visit us at http://www.nuklearmalaysia.org

With Best Compliments

ADVANCED NUCLEAR ENGINEERING RESEARCH GROUP (UTM)

NUCLEAR ENGINEERING STUDENTS SOCIETY (UTM)

ASME MALAYSIA SECTION

MALAYSIAN WELDING AND JOINING SOCIETY

IOP Conf. Series: Materials Science and Engineering 1231 (2022) 011002 doi:10.1088/1757-899X/1231/1/011002

Peer Review Statement

All papers published in this volume have been reviewed through processes administered by the Editors. Reviews were conducted by expert referees to the professional and scientific standards expected of a proceedings journal published by IOP Publishing.

- **Type of peer review:** Single Anonymous
- **Conference submission management system:** Morressier
- **Number of submissions received:** 34
- **Number of submissions sent for review:** 24
- **Number of submissions accepted:** 22
- **Acceptance Rate (Submissions Accepted / Submissions Received × 100):** 64.7
- **Average number of reviews per paper:** 2
- **Total number of reviewers involved:** 46
- **Contact person for queries:**
  Name: Faridah Bt Mohd Idris
  Email: faridah_idris@nuclearmalaysia.gov.my
  Affiliation: Dr

IOP Conf. Series: Materials Science and Engineering 1231 (2022) 012001 doi:10.1088/1757-899X/1231/1/012001

Model predictive and fuzzy logic controllers for reactor power control at Reaktor TRIGA PUSPATI

Mohd Sabri Minhat[1,2], Nurul Adilla Mohd Subha[1, a)], Fazilah Hassan[1], Abdul Rashid Husain[1], Anita Ahmad[1]

[1]School of Electrical Engineering, Faculty of Engineering, Universiti Teknologi Malaysia, 81310 Skudai, Malaysia.
[2]Reactor Instrumentation and Control Engineering, Reactor Technology Centre, Malaysian Nuclear Agency, 43000 Kajang, Malaysia

[a)]nuruladilla@utm.my

Abstract. The Reaktor TRIGA PUSPATI (RTP) rely extensively on a core power control system for control reactivity to fulfilment the fundamental safety function for a nuclear reactor. It is technically challenging to operate within tight multiple parameter constraints and keep the core power output stable. At present, the power tracking performance of the system could be considered unsatisfactory which produces a relatively long settling time and high control effort. Hence, a study of a model predictive control (MPC) strategy by integrating the current Power Change Rate Constraint (PCRC) using fuzzy logic which is part of the core power control design is conducted. In this paper, the MPC design based on mathematical models of the reactor core included point kinetics model, thermal-hydraulic model, reactivity model and dynamic rod position model help to enhance core power control. The power tracking performance of the proposed control method and previous Feedback Control Algorithm-Fuzzy PCRC is compared via computer simulation with different power range operations. Overall, the results show the MPC-Fuzzy PCRC strategy provides a greater level of operational safety and optimum operation of the TRIGA reactor.

1. Introduction

The Reaktor TRIGA PUSPATI (RTP), TRIGA MARK II type is the only nuclear research reactor available in Malaysia. The reactor is used to generate neutrons for various research purposes such as medical, material study, and industrial applications. The generation of thermal power or core power is varied based on the movement of the control rods and can be regulated by the core power control system.

In general, the automatic core power control is a part of the Instrumentation and Control (I&C) system which is designed to provide efficient and safe power production [1]. At present, the tracking performance of the power control system at RTP is deemed unsatisfactory due to slow tracking, unsmooth transient response, and a long settling time. As a result, continuous improvement is still required for developing a stable and safe core power control system.

The core power control is physically designed based on common safety parameter constraints such as step reactivity limiting [2], reactor period limiting [3], power change rate-limiting [4], and control rod speed limiting [5]. To date, the effects of the combination of all these safety nuclear reactor parameter constraints in the other type of controller which is capable of naturally handling constraints of the plant have not been extensively studied. Generally, control rod speed or velocity constraint is the most popular safety constraint which is much easier to design for implementation in a practical system. However, in TRIGA research reactor is require tight multiple parameter constraints to regulate reactor power [6].

The RTP uses a conventional core power control known as Feedback Control Algorithm (FCA) [7]; [8] for the power manoeuvring up to 1 MWth based on the conventional Control Rod Velocity Design (cCRVD) and conventional Power Change Rate Constraint (cPCRC). In our recent works, the new PCRC based on the fuzzy logic approach [9] and changes in the maximum rod speed limiter values [10] have been designed to provide better results than the FCA-conventional PCRC method in terms of reducing chattering error. However, the result produced is not fully optimized in terms of settling and rise time. Hence, in this paper, the model-based controller using Model Predictive Control (MPC) and combination Fuzzy PCRC of the strategies in [9] is studied and analyzed to further improve the power tracking performance of the RTP.

Most studies using analytical for designing the MPC core power control as the current practices [11], [12], and other possibilities methods such as System Identification (ID) help to simplify the process and convert the non-linear model based on zero dimension approach in ordinary differential equation (ODE) to state-space form. In this study, to evaluate the effectiveness of the developed MPC and Fuzzy PCRC strategies, a set of performance criteria based on a descriptive approach which is most commonly used in this core power control application area such as settling time, rise time, per cent overshoot, chattering error, offset error, workload, and energy released.

This paper is organized as follows. The modelling of RTP using system identification is presented in Section 2 and the current FCA-Fuzzy PCRC RTP core power control system is briefly described in Section 3. The proposed model predictive control strategy is presented in Section 4. The results and discussion on the implementation of an MPC-Fuzzy PCRC strategy are given in Section 5. Finally, conclusions are given at the end of the paper.

2. Reaktor TRIGA PUSPATI model using System Identification

The TRIGA model is based on point kinetics, and it includes six groups of delayed neutrons, as well as reactivity feedback from control rod movement, fuel temperature, and moderator coolant temperature changes [13].

The linearized model, which is extensively used in RTP modelling can be expressed in state-space form as [14][15]:

$$\left.\begin{aligned}
\dot{x} &= Ax + Bu \\
y &= Cx + Du \\
A &= \begin{bmatrix}
-\frac{\beta}{\Lambda} & \lambda_1 & \lambda_2 & \lambda_3 & \lambda_4 & \lambda_5 & \lambda_6 & \frac{\alpha_f}{\Lambda}\psi_0 & \frac{\alpha_m}{\Lambda}\psi_0 & \frac{\psi_0}{\Lambda} \\
\frac{\beta_1}{\Lambda} & -\lambda_1 & 0 & 0 & 0 & 0 & 0 & 0 & 0 & 0 \\
\frac{\beta_2}{\Lambda} & 0 & -\lambda_2 & 0 & 0 & 0 & 0 & 0 & 0 & 0 \\
\frac{\beta_3}{\Lambda} & 0 & 0 & -\lambda_3 & 0 & 0 & 0 & 0 & 0 & 0 \\
\frac{\beta_4}{\Lambda} & 0 & 0 & 0 & -\lambda_4 & 0 & 0 & 0 & 0 & 0 \\
\frac{\beta_5}{\Lambda} & 0 & 0 & 0 & 0 & -\lambda_5 & 0 & 0 & 0 & 0 \\
\frac{\beta_6}{\Lambda} & 0 & 0 & 0 & 0 & 0 & -\lambda_6 & 0 & 0 & 0 \\
\frac{N_0}{\mu_f} & 0 & 0 & 0 & 0 & 0 & 0 & -\frac{\Omega}{\mu_f} & \frac{\Omega}{\mu_f} & 0 \\
0 & 0 & 0 & 0 & 0 & 0 & 0 & \frac{\Omega}{\mu_m} & \frac{(\Omega+2M)}{\mu_m} & 0 \\
0 & 0 & 0 & 0 & 0 & 0 & 0 & 0 & 0 & 0
\end{bmatrix} \\
B &= [0 \quad 0 \quad 0 \quad 0 \quad 0 \quad 0 \quad 0 \quad 0 \quad 0 \quad G_r]^T \\
C &= [1 \quad 0 \quad 0 \quad 0 \quad 0 \quad 0 \quad 0 \quad 0 \quad 0 \quad 0] \\
D &= [0]
\end{aligned}\right\} \quad (1)$$

where A is the state matrix, B is the input matrix, C is the output matrix, D is the disturbance matrix, core state $x = [\delta N \; \delta\eta_1 \; \delta\eta_2 \; \delta\eta_3 \; \delta\eta_4 \; \delta\eta_5 \; \delta\eta_6 \; \delta T_m \; \delta T_f \; \delta\rho]^T$, core input $u = [z_r]$, and core output $y = [N]$.

All the parameters used in this paper are summarized in Table 1.

Table 1. Nomenclature.

ψ	Relative neutron density	Λ	Mean neutron generation time (s)
ψ_o	Relative neutron density at initial equilibrium state	β	The total fraction of effective delayed neutron (Δk/k)
G_r	Reactivity worth of the control rod (*Δk/k/m*)	β_i	The *i*-th group of the delayed neutron (*Δk/k*)
N	Actual core power (*W*)	α_f	Reactivity due to change in temperature fuel (*Δk/k/°C*)
N_o	Nominal core power (*W*)	α_m	Reactivity due to change in temperature moderator (*Δk/k/°C*)
μ_m	Heat capacity of coolant (Ws°C^{-1})	η_i	The *i*-th group of normalized precursor concentration (m^{-3})
μ_f	Heat capacity of fuel (Ws°C^{-1})	T_f	The average temperature of the fuel (°*C*)
Ω	Global heat transfer coefficient ($W\,^0C^{-1}$)	T_m	The average temperature of coolant (°*C*)
λ_i	Decay constant of the *i*-th group of delay neutron precursor (s^{-1})	ρ	Total reactivity
M	Heat capacity of a mass flow rate of coolant ($W\,^0C^{-1}$)	z_r	The velocity of the control rod bank (ms^{-1})

In order to have high consistency and accuracy in the RTP model included the control rod position dynamic in [8] can also be derived based on a set of actual input and output data using System ID. The plant model's linearized state-space model based on System ID can be represented as [16]

$$\begin{cases} x(t+t_s) = A_{SID}x(t) + B_{SID}u(t) + K_{SID}e(t) \\ y(t) = C_{SID}x(t) + D_{SID}u(t) + e(t) \end{cases} \tag{2}$$

where A_{SID}, B_{SID}, C_{SID} and D_{SID} are coefficient matrices estimate using System ID, K_{SID} represents the noise matrix of the model and $e(t)$ is the disturbance that exists in the model output.

Based on the measurement of input, which is control rod velocity values, and output, which is core power data, a mathematical model of the RTP is built. A square wave signal is employed as a perturbed input to the system in this study, and 2401 observations of input and output data sets from a non-linear model simulation were collected at 0.5 seconds. The square wave signal is used as a representation of the output from the controller in automatic operation mode and is still capped at a maximum value of 23 steps/cycle and a minimum value of 0 step/cycle.

The measured input and output data are divided into two sets in System ID; the first set is used for estimation, while the second set is used for validation. The first 819 samples of data were used for estimation, while the remaining samples were used for validation. The System ID Toolbox in the Matlab environment is used to determine a suitable model structure for the RTP. In Matlab System ID Toolbox, there are a few model structures that are often utilised in real-world applications. Model-based Predictive Control (MPC) state-space model identification with 12th model order, 0.5 s model sample time (t_s), and state-space core model structure is employed in this study.

Model validation is the last step in the System ID process. The second set of data, sampled from 1583 to 2401, will be utilised for validation reasons, as stated at the beginning of this subsection. Model validation evaluates the discrepancy between actual and simulated data to determine whether the identified model appropriately describes the process under investigation. If the best fit is more than 95%, the created model is acceptable. Eq. (3) can be used to calculate the best fit.

$$Best\ Fit = \left(1 - \frac{|y - \hat{y}|}{|y - \bar{y}|}\right) \text{x } 100 \tag{3}$$

where y is the actual measure output, $\hat{y}$ is simulated or predicted model output and $\bar{y}$ is the mean of y. 100% corresponds to a perfect fit, and 0% indicated that the fit is no better than guessing the output to be a constant where $\hat{y} = \bar{y}$.

The best fit result for the RTP model using System ID is illustrated in Figure 1.

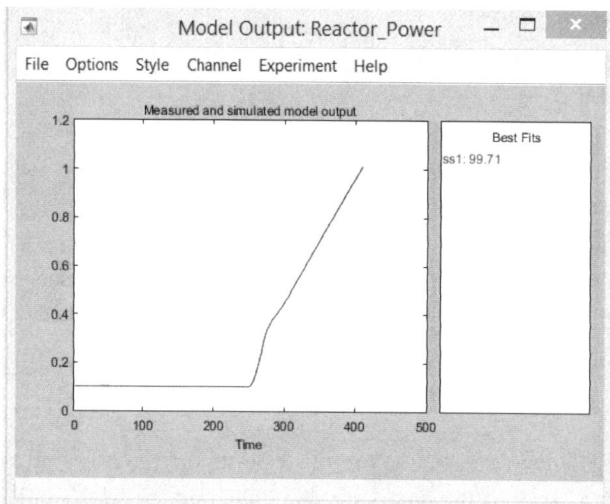

Figure 1. Development of state-space RTP model via System ID in Matlab

3. Feedback Control Algorithm-Fuzzy PCRC core power control system

The RTP feedback core power control system for power manoeuvring is illustrated in Figure 2. The term input refers to a Power Demand (PDM) and the output is the neutron power at the core. The core power is measured by an ex-core neutron detector and Neutron Measurement System (NMS) as signal processing. The NMS provides two signals which are core power (*N*) and the rate of power change (Log Rate). The error deviation in percentage between the PDM and the core power output is used as the inputs for the signal filter and Proportional-Integral (PI) controller. The controller output in form of the control rod velocity is fed to the Control Rod Velocity Design (CRVD) to constraint the reactivity insertion rate in the core. The Fuzzy PCRC is used to penalize control rod velocity signal using a different gain during a specific power level condition before entering to Control Rod Drive Mechanism (CRDM). Using the said configuration, the core power control has a 1.25% full power (FP) chattering error with a large settling time in the case of a sudden change in power demand.

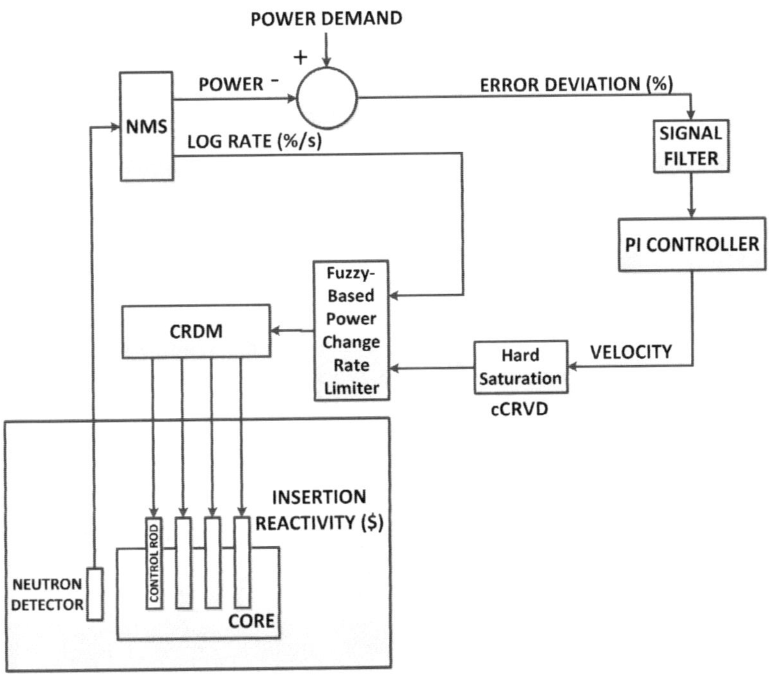

Figure 2. Block diagram of core power control with Fuzzy PCRC

The FCA consist of PCRC, signal filter, PI controller, cCRVD, and CRDM. The PI controller with a signal filter is designed as follows [9]; [15]:

$$\left.\begin{aligned} u_c &= K_P E_{fi} + K_I \int_{t_0}^{t} E_{fi} dt \\ E &= \left[G_1 log\left(\frac{PDM}{N}\right)\right]_{\pm 1} \\ E_{fi}(k) &= 1.47197E(k) + 0.882\left[E_{fi}(k-1) - E(k-1)\right] \end{aligned}\right\} \quad (4)$$

where u_c is the output signal from the controller, G_1, K_P, K_I, are controller tuning gain for FCA, and E_{fi} is input filter calculation based on the error signal (E).

In fuzzy logic rules, the most commonly used parameterized membership functions (MFs) are triangular, trapezoids, bell curves, Gaussian and sigmoidal functions [17]. However, this study only considered a single type of MFs which is triangular.

In this study, the range of rate of power change need to specify for each condition; very fast (more than u_c), fast (u_c), slow (u_b), very slow (u_{ab}) and no change (u_a), The rules with triangular MFs are shown in Figure 3.

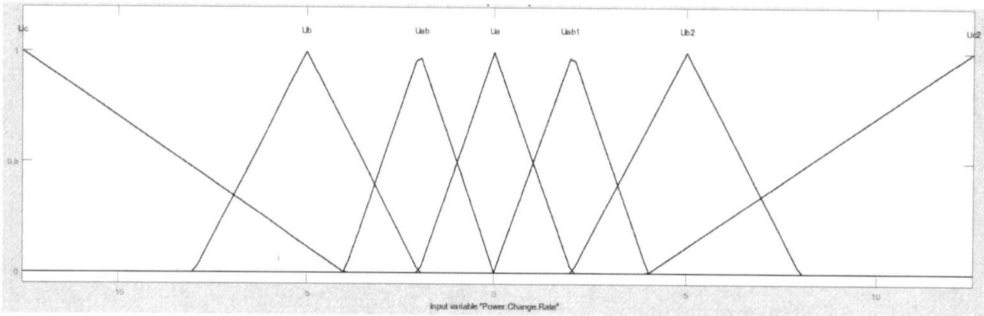

Figure 3. The triangular MFs for the Fuzzy PCRC.

Before applying the fuzzy operator, the rule weight is first assigned. Each rule has weight, w_i can vary from 0 to 1.0, which is applied to each part of the antecedent or single fuzzy degree of membership. Based on Eom et al., by setting w_i is 1.0 for the purpose to provide maximum penalize value in order to stop control rod moving and w_i is 0 for introducing not constraint effect on the control rod velocity value calculated by the controller. However, there is no rule of thumb to decide the value of w_i and the number of fuzzy rules needed for each power increment. For the case of RTP, the chattering error at steady-state maybe can be reduced by the fine movement of control rod velocity when near to power demand by introducing small weight without zero value. Thus, in this study, the w_i value is changed to smaller than 1.0 which can decrease the sensitivity on the system and effect of one rule relative to the others. At this stage, the PCRC is limited to ±12.5% FP/s.

The proposed value of w_i has been assigned to each rule based on a set of inputs consisting of the rate of power change with their corresponding output values in [9] for the triangular MFs Fuzzy PCRC. The rules can be expressed as

$$\left.\begin{array}{r} Rule\ 1.\ If\ (Power.Change.Rate\ is\ u_c)\ then\ (output\ is\ V_s^{c})\ (w_c = 1.0) \\ Rule\ 2.\ If\ (Power.Change.Rate\ is\ u_b)\ then\ (output\ is\ V_s^{b})\ (w_b = 0.15) \\ Rule\ 3.\ If\ (Power.Change.Rate\ is\ u_{ab})\ then\ (output\ is\ V_s^{ab})\ (w_{ab} = 0.01) \\ Rule\ 4.\ If\ (Power.Change.Rate\ is\ u_a)\ then\ (output\ is\ V_s^{a})\ (w_a = 0.001) \\ Rule\ 5.\ If\ (Power.Change.Rate\ is\ u_{ab1})\ then\ (output\ is\ V_s^{ab})\ (w_{ab} = 0.01) \\ Rule\ 6.\ If\ (Power.Change.Rate\ is\ u_{b2})\ then\ (output\ is\ V_s^{b})\ (w_b = 0.15) \\ Rule\ 7.\ If\ (Power.Change.Rate\ is\ u_{c2})\ then\ (output\ is\ V_s^{c})\ (w_c = 1.0) \end{array}\right\} \qquad (5)$$

In addition, the following scaled control inputs to relate between the rule weight and control rod velocity calculated by the controller are considered:

$$V_s^i = w_i u_{step} \qquad (6)$$

where variable w_i (for $i = a, b, c, ab$) is an adjustable weighting parameter and can be varied at different levels of the penalty based on the rate of power change.

4. TRIGA model predictive core power control

In general, the MPC controller performs all estimation and optimization calculations using a discrete-time. Therefore, the discrete form of the RTP as presented previously in Eq. **Error! Reference source not found.** can be expressed as:

$$\begin{cases} x(k+1) = Ax(k) + B_u u(k) + B_{v_o} v_o(k) + B_d d(k) \\ y(k) = Cx(k) + D_{v_o} v_o(k) + D_d d(k) \end{cases} \quad (7)$$

where k is the time index; x is the state of the model, u is the vector of manipulated variables (control rod velocity), v_o is the vector of measured disturbances, d is the vector of unmeasured disturbances, and y is the output vector (core power). Whereas A, B, C, and D are the constant state-space matrices.

The model used for both prediction and state estimation is augmented with the disturbance $d(k)$ to ensure zero offset tracking in the steady-state. However, for this study, the designed MPC did not consider measured and unmeasured disturbance inputs by setting $B_{v_o}(k)$, $B_d(k)$, $D_{v_o}(k)$, and $D_d(k)$ to zero. Then, Eq. (7) can be rewritten as:

$$\begin{cases} x(k+1) = Ax(k) + Bu(k) \\ y(k+1) = Cx(k+1) + Du(k) \end{cases} \quad (8)$$

where the variables are considered as the deviations from the nominal values based on the operating zone where the model is defined.

In general, the controlled quality is defined as the velocity of the control rod that can be considered as $u = 0$ when the actual core power is stable at the desired power level. Thus, the objective cost function J for the MPC system is defined as [18]:

$$J = (R_s - Y)^T (R_s - Y) + U^T R_W U \quad (9)$$

where $Y = [y(k+1) \quad y(k+2) \quad \cdots \quad y(k+N_p)]^T$; $R_s = [0.75 \quad 0.75 \quad \cdots \quad 0.75]^T r(k)$; r is the reference trajectory (power demand) of core power; $U = [u(k) \quad u(k+1) \quad \cdots \quad u(k+N_c-1)]^T$; N_p is the prediction horizon; N_c is the control horizon; $R_W = R_1 \begin{bmatrix} 0.75 & 0 & \cdots & 0 \\ 0 & 0.75 & \cdots & 0 \\ \vdots & \vdots & \ddots & 0 \\ 0 & 0 & 0 & 0.75 \end{bmatrix}$, R_W is the weight matrix with N_c x N_c dimensions and R_1 is a tuning parameter for the desired closed-loop performance.

Defining $\emptyset = \begin{bmatrix} CB & 0 & 0 & \cdots & 0 \\ CAB & CB & 0 & \cdots & 0 \\ CA^2B & CAB & CB & \cdots & 0 \\ \vdots & \vdots & \vdots & \ddots & \vdots \\ CA^{N_p-1}B & CA^{N_p-2}B & CA^{N_p-3}B & \cdots & CA^{N_p-N_c}B \end{bmatrix}$ and $F = \begin{bmatrix} CA \\ CA^2 \\ \vdots \\ CA^{N_p} \end{bmatrix}$ gives:

$$Y = Fx(k) + \emptyset U \quad (10)$$

To find the optimal solution, the derivative of J with respect to U is needed. After $\frac{\partial J}{\partial U} = 0$ is defined, the optimal solution of the control system is defined as:

$$U = (\emptyset^T \emptyset + R_W)^{-1} \emptyset^T (R_s - Fx(k)) \quad (11)$$

As the movement of the control rod is mechanically limited, the limit on $u(k)$ is existent. The maximum velocity of the control rod is set to 23 steps per cycle, and the height of the reactor core is about 0.381 m. It is about 21,525 steps for the control rod to move from the bottom to the top of the reactor core. Consequently, the range of $u(k)$ is obtained:

The velocity of the control rod, $u(k)$ = 23 step per cycle x 0.0177 mm per step.

$$-0.4 \leq u(k) \leq 0.4 \tag{12}$$

Since the limits on $u(k)$ are existent, quadratic programming (QP) is utilized for optimization. Another objective cost function J_{QP} is defined as [19]:

$$J_{QP} = \frac{1}{2}U^T(\emptyset^T\emptyset + R_W) + \left(\emptyset^T\left(-R_s + Fx(k)\right)\right)^T U \tag{13}$$

where J_{QP} is qualitatively equivalent to J for the MPC system.

Defining control rod velocity constraint gives:

$$\begin{cases} -A_{QP}U \leq M \\ A_{QP}U \leq M \end{cases} \tag{14}$$

where $M = [0.4 \quad 0.4 \quad \cdots \quad 0.4]^T$ and A_{QP} is a unit matrix with N_c x N_c dimensions.

In this section, the previous Fuzzy PCRC proposed in [9] is integrated with the general MPC algorithm in Eq. (7) – (14) and later will be compared with Fuzzy PCRC with FCA in Section 5.

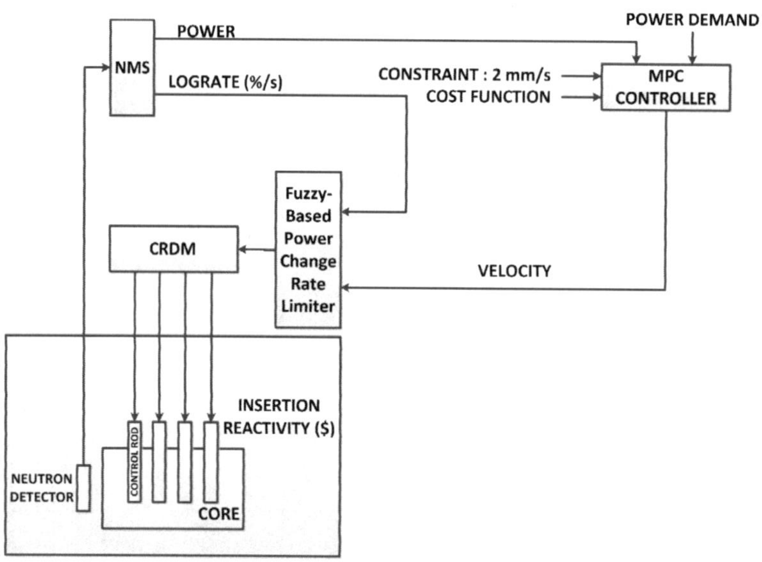

Figure 4. The block diagram of the MPC with triangular MFs Fuzzy PCRC.

To ensure the simplicity of the developed function, the MPC-Fuzzy PCRC in Figure 4 is designed using the same approach as FCA-Fuzzy PCRC. The triangular MFs MPC-Fuzzy PCRC is designed as follows:

$$\breve{u}_{PCRC} = \left[1 - \left(\frac{u_a w_a + u_b w_b + u_c w_c + u_{ab} w_{ab}}{u_a + u_b + u_c + u_{ab}}\right)\right] u_{MPC} \quad (15)$$

where variable u_i (for $i = a, b, c, ab$) is fuzzy set for triangular MFs, variable w_i (for $i = a, b, c, ab$) is adjustable weighting parameter, and u_{MPC} is the control rod velocity signal calculated using the MPC controller.

5. Results and discussion

The linearized model of the RTP is modelled using System ID Eq. (2) presented in Section 2. The validation of the RTP simulation model using the experimental data has been reported in our previous work [8]; [10]. The control horizon is determined by the rule of thumb in the Matlab simulation manual [20], in which the control horizon is set to 10% to 20% of the prediction horizon and has a minimum of 2 to 3 steps. In this study, the control horizon is set to 5 steps and each step is equal to 0.2 s, thus the interval time is equal to 1 s, by considering the time delay produced by the motor. Based on the considered rule of thumb, the prediction horizon is chosen to be within 10 to 200 steps. Within the selected range, the optimum prediction horizon is determined by testing all the prediction horizon values. For simplicity, the prediction values are divided into 2 groups. The first group is 10 steps to 80 steps in the range of 6% to 50% to test the rule of thumb in Matlab simulation manual and the second group is between 90 to 200 steps in the range of 2% to 5% to test beyond the rule of thumb in the first group. The considered prediction values are not more than 200 steps, equal to 40 s because it is not suitable for reactor operation to have tool long prediction values, which consequently will produce a long settling time. The result of power tracking response and actuation signal based on these two groups are shown in Figure 5 and Figure 6.

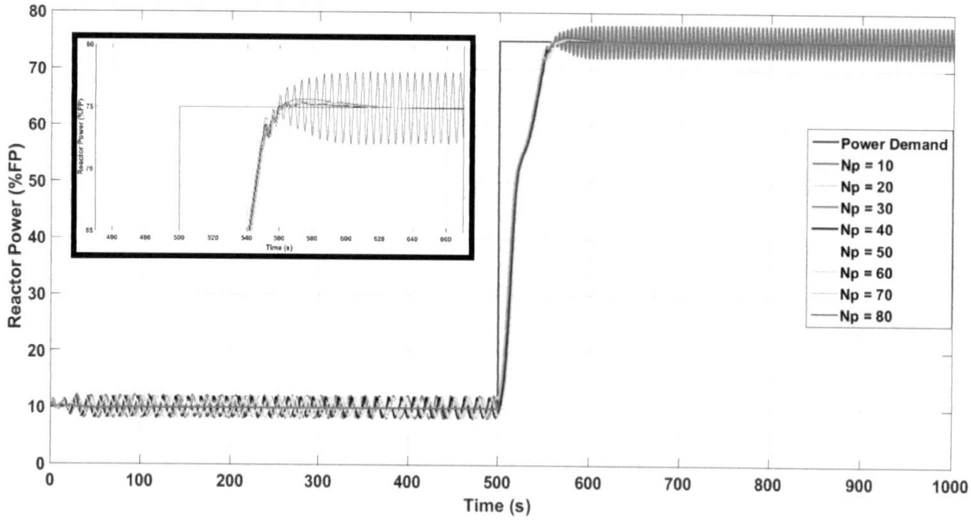

(a)

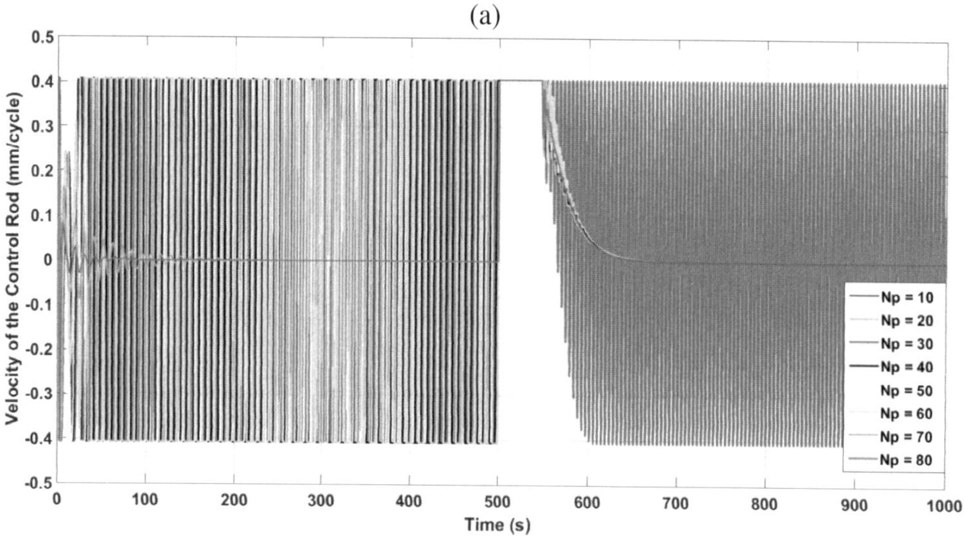

(b)

Figure 5. (a) The power tracking performance for prediction horizon 10 to 80, (b) The velocity control signal for prediction horizon 10 to 80

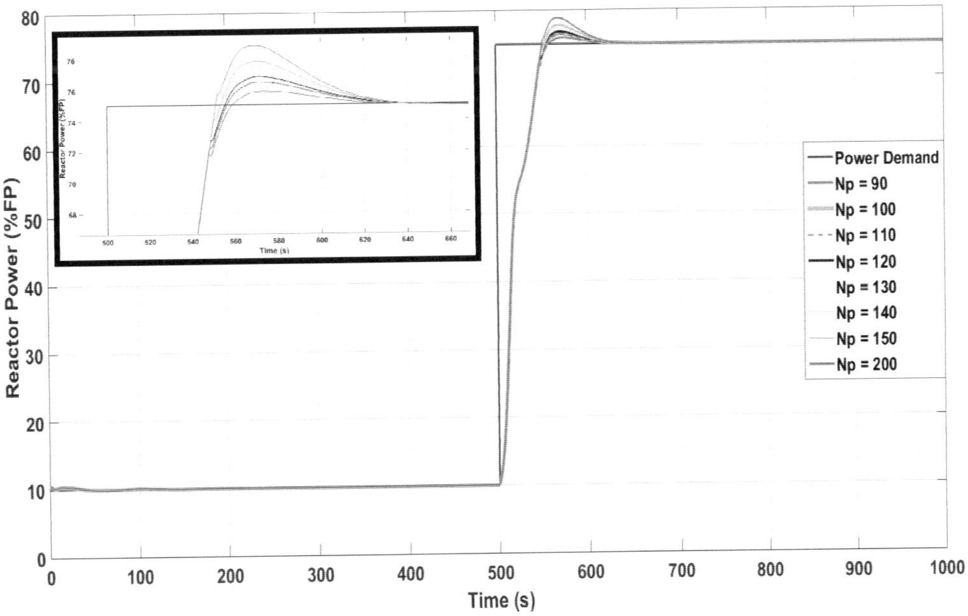

Figure 6 (a)

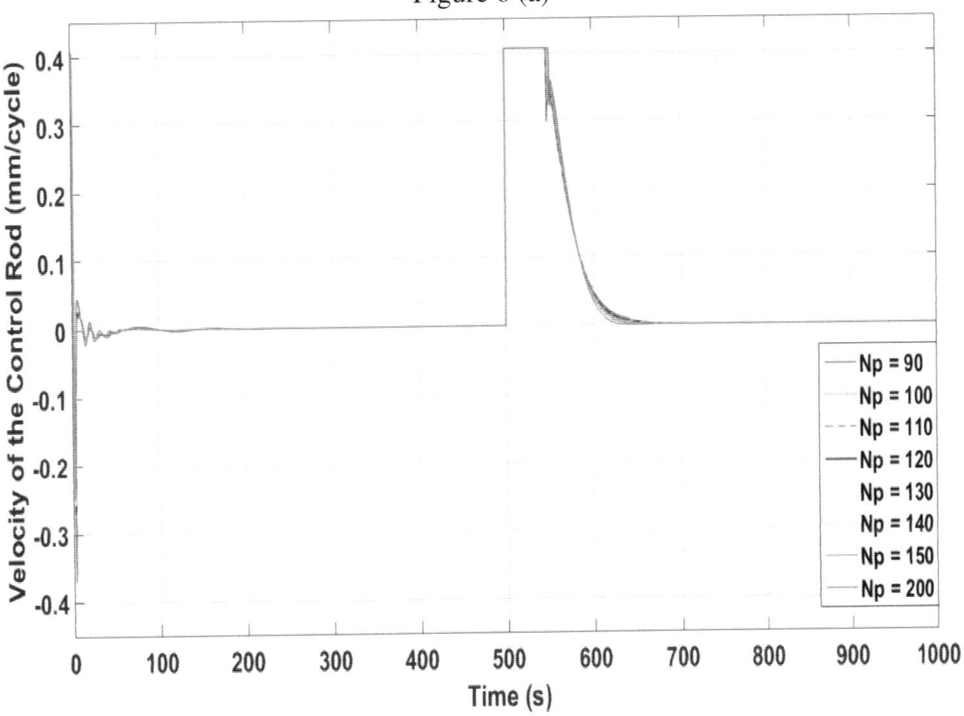

Figure 6 (b)

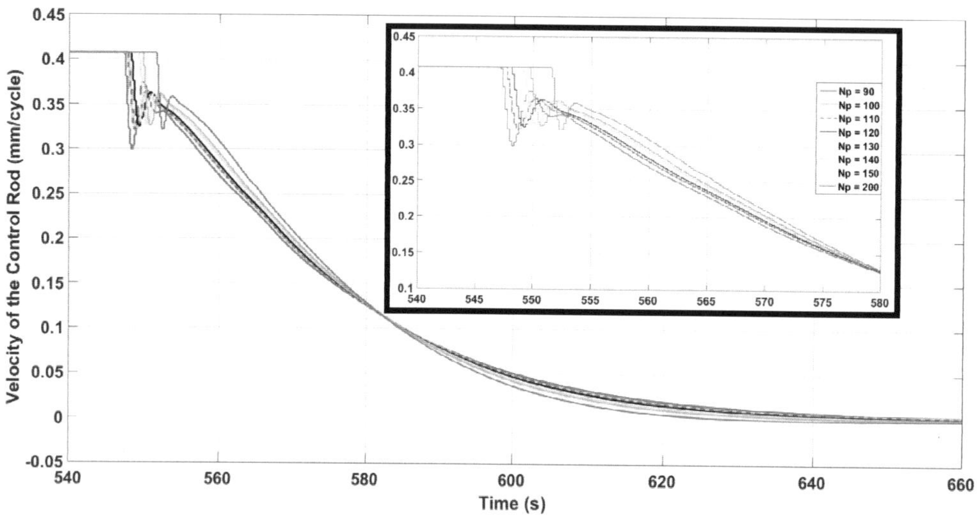

Figure 6 (c)

Figure 6. (a) The power tracking performance for prediction horizon 90 to 200, (b) The velocity control signal for prediction horizon 90 to 200, (c) Closed view of the actuation signal for prediction horizon 90 to 200

Small values of the prediction horizon in the first group as shown in Figure 5 tend to produce oscillations in the control rod movement and the power output, thus it is not suitable for safe operation in a reactor. The prediction horizon above 70, however, produces minimal oscillation in both signals and disappears when the prediction horizon reaches above 90. All prediction horizon values in the second group have small power overshoot, in which the maximum overshoot is around 6.7% which is less than 10% for safety requirements. Based on the result in Figure 6, this study chose the prediction horizon equal to 100, by taking into consideration that the value complies with the minimum criterion for the safe operation in core power control, which are minimum overshoot, less oscillation, and continuous increase reactor power.

The MPC parameters are set to 100 steps of prediction horizon N_p, which equal 20 seconds, 5 steps control horizon N_c, to update the control signal for every 1 second, and $R_1 = 0.1$. The performance of the MPC is compared with FCA-Fuzzy PCRC through the power manoeuvring of the core power and actuation signal of CRDM. Figure 7 shows that the MPC control strategy can significantly improve the power tracking performance compared to the FCA-Fuzzy PCRC during a transient. However, when core power is a power above 70% FP, all MPC produced small ripple and overshoot. In terms of smoothness of the actuation signal, the FCA-Fuzzy PCRC provide a better signal to CRDM. The signal fluctuation occurred when MPC reduce the control rod velocity from a maximum value to zero due to a mismatch in model prediction. The MPC struggled to eliminate the prediction error, thus producing higher control effort and consequent increase in the chattering error.

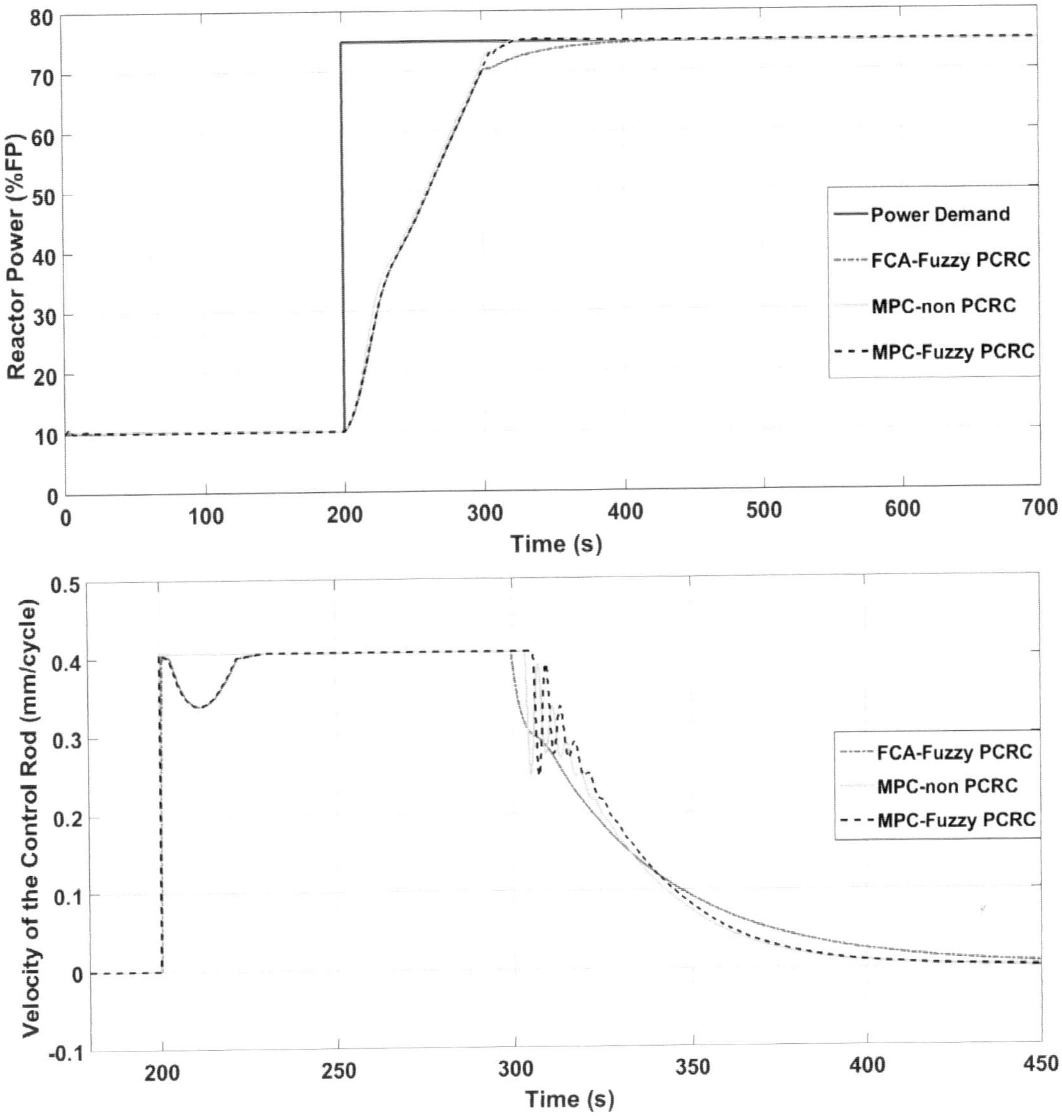

Figure 7. Comparison of different controller types for limiting the rate of power increment by increasing the power to 75% FP

The final cost value is calculated using the linear MPC with 1.220e-7 respectively. Both MPC controllers have achieved their objectives by minimizing cost value to reduce the control effort. The quantitative performance comparison is in Table 2.

Table 2. Transient and steady-state response for FCA and MPC-Fuzzy PCRC core power control at 750kW

Type	*Settling Time, T_s (s)*	*Rise Time, T_r (s)*	*Percent Overshoot, P_{os} (%)*	*Chattering Error during transient, Δe_{ce} (%)*	*Chattering Error during steady-state, Δe_{ce} (%)*	*Offset Error, Δe_{oe} (%)*	*Work Load (mm/cycle) / Energy Released (kW-h)*
FCA-Fuzzy PCRC	113.5	86.0	0.00131	0.02621	0.026213	+0.00098	105.55 / 5299.06
MPC-non PCRC	110.0	84.5	0.32777	0.31427	0.000468	+0.01397	105.85 / 5391.29
New MPC-Fuzzy PCRC	110.5	85.5	0.32814	0.31439	0.000469	+0.01422	105.85 / 5352.49

By referring to Table 2, the MPC-non PCRC type provides the most responsive reactivity compared to other control strategies with the reduction in settling and rise times. In addition, it is also capable of optimizing total energy released from the reactor core significantly. The combination of Fuzzy PCRC and MPC type does not produce any significant improvement compared to the MPC-non PCRC. However, the effect of input-output fuzzy rules is maintained to improve the chattering error when different controller type is used. The absent PCRC in MPC caused the control movement to maintain maximum control rod velocity. Overall, based on Table 2, the MPC is the best control strategy in terms of reducing settling and rise time. However, overshoot, smoothness of the actuation signal and offset error still need to be improved for MPC to be chosen. The FCA is the best solution to eliminate the offset error during steady-state.

The MPC-non PCRC and MPC-Fuzzy PCRC are also tested with 3 mm/s control rod velocity and highest reactivity insertion rate by using Regulating (RG) rod. It can be observed that the rate of power change for the MPC-non PCRC is more than 12.5% FP/s. By increasing rod velocity, can significantly improve tracking performance for wide-range power demand. The smoothness of the actuation signal can be observed by using MPC-Fuzzy PCRC. However, without the PCRC component, the tracking performance considered is not safe in terms of the rate of power change which is not a bounded constraint and more than the currently permitted 12.5% FP/s during a transient as shown in Figure 8. Therefore, MPC-Fuzzy PCRC is introduced for high control rod velocity and reactivity insertion rate.

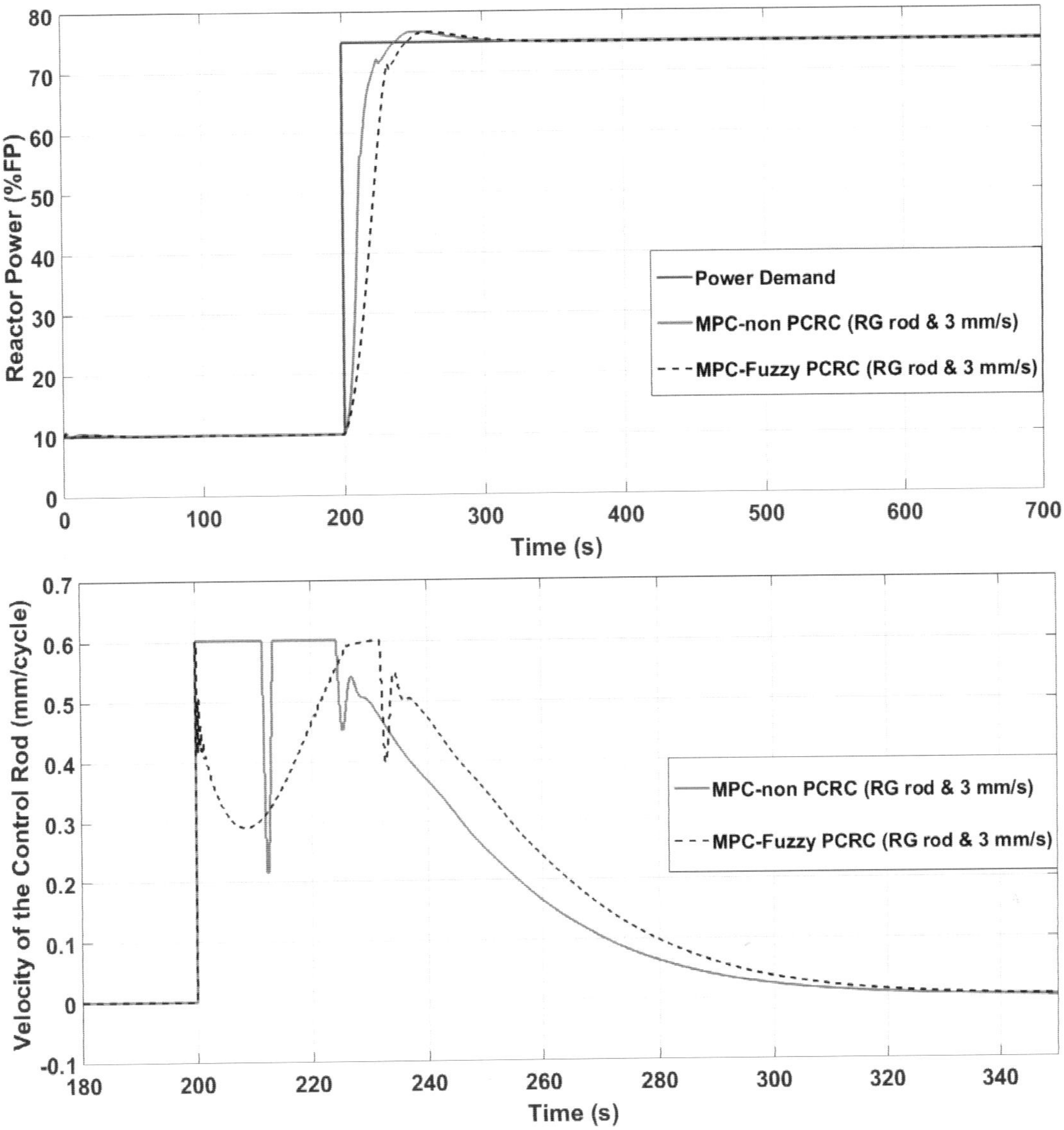

Figure 8. The power tracking performance and velocity control signal between MPC-non PCRC and MPC-Fuzzy PCRC using RG rod at 3 mm/s rod velocity

6. Conclusion

A study of the MPC-Fuzzy PCRC core power control strategies for the TRIGA reactor is presented to provide a greater level of operational safety and is very useful for types of nuclear that require tight multiple parameter constraints to regulate reactor power. Instead of using FCA-Fuzzy PCRC, the new MPC-Fuzzy PCRC control strategies consider other alternative solutions to optimize the TRIGA core power control performance. Overall, the results show that the response from the MPC-Fuzzy PCRC offers better results than the FCA-Fuzzy PCRC, which reduces the chattering error during steady-state by up to 98%, the settling time by up to 2.6%, and the rise time up to 0.6%. The optimised energy released from the reactor core is slightly improved by 1.0% for the same workload with MPC-Fuzzy PCRC. The significant finding from this study is that the MPC and Fuzzy can mutual understand and collaborate for different functions assigned. The MPC controller can provide a fast response due to model prediction ability and fuzzy logic to penalize control rod velocity signal based on power change rate to improve the drivability of the CRDM.

7. References

[1] Ikonomopoulos A, Varvayanni M and Catsaros N 2015 Instrumentation and control implementations in research reactors : A review *Int. Conf. Nucl. Energy New Eur.* 1–9.
[2] Benítez-Read J S, Ruan D, Najera-Hernandez M, Perez-Clavel B, Pacheco-Sotelo J O and Lopez-Callejas R 2005 Comparison between a continuous and discrete method for the aggregation and deffuzification stages of a TRIGA reactor power fuzzy controller *Prog. Nucl. Energy* **46** 309–320.
[3] Pérez-Cruz J H and Poznyak A 2008 Neural control for power ascent of a TRIGA reactor *2008 American Control Conf.* (Seattle, WA, USA: IEEE) 2190–2195.
[4] Eom M, Chwa D and Baang D 2015 Robust Disturbance Observer-Based Feedback Linearization Control for a Research Reactor Considering a Power Change Rate Constraint *IEEE Trans. Nucl. Sci.* **62** 1301–1312.
[5] Kim J H, Park S H and Na M G 2014 Design of a model predictive load-following controller by discrete optimization of control rod speed for PWRs *Ann. Nucl. Energy* **71** 343–351.
[6] Malaysian Nuclear Agency 2020 *Safety analysis report for Reaktor TRIGA PUSPATI*
[7] Minhat M S, Selamat H and Mohd Subha N A 2018 Adaptive control method for core power control in TRIGA Mark II reactor *IOP Conf. Ser. Mater. Sci. Eng.* **298** 12028.
[8] Minhat M S, Mohd Subha N A, Hassan F and Mohamad Nordin N 2020 An improved control rod selection algorithm for core power control at TRIGA PUSPATI Reactor *J. Mech. Eng. Sci.* **14** 6362–6379.
[9] Minhat M S, Adilla N, Subha M and Hassan F 2020 Application of fuzzy logic for power change rate constraint in core power control at Reaktor TRIGA PUSPATI *IOP Conf. Ser. Mater. Sci. Eng.* **785** 12022.
[10] Minhat M S, Adilla N, Subha M, Hassan F, Ahmad A and Rashid A 2020 Profiling and analysis of control rod speed design on core power control for TRIGA reactor *Prog. Nucl. Energy* **128** 103481.
[11] Etchepareborda A and Lolich J 2007 Research reactor power controller design using an output feedback nonlinear receding horizon control method *Nucl. Eng. Des.* **237** 268–276.
[12] Andraws M S, Abd El-Hamid A A, Yousef A H, Mahmoud I I and Hammad S A 2017 Performance of Receding Horizon Predictive Controller for Research Reactor *12th Int. Conf. on Comp. Eng. and Sys. (ICCES)* (Cairo, Egypt: IEEE) 272–278.
[13] Cammi A, Ponciroli R, Borio A, Magrotti G, Prata M, Chiesa D and Previtali E 2013 A zero dimensional model for simulation of TRIGA Mark II dynamic response *Prog. Nucl. Energy* **68** 43–54
[14] Shaffer R A, Edwards R M and Lee K Y 2005 Design and validation of robust and autonomous control for nuclear reactors *Nucl. Eng. Technol.* **37** 139–150.
[15] Minhat M S, Subha N A M, Hassan F, Husain A R, Ahmad A, Ismail F S and Hamzah N 2021 A multipronged core power control strategy for Reaktor TRIGA PUSPATI *IOP Conf. Ser. Mater. Sci. Eng.* **1106** 12001.
[16] Ljung L 2021 *System Identification Toolbox Gettting Started Guide* (The MathWorks)
[17] Wai T C 2009 *Fuzzy associative memory architecture* (Nanyang Technological Univ.).
[18] Wang G, Wu J, Zeng B, Xu Z, Wu W and Ma X 2017 Design of a model predictive control method for load tracking in nuclear power plants *Prog. Nucl. Energy* **101** 260–269.
[19] Wang G, Wu J, Zeng B, Xu Z and Wu W 2016 State-Space model predictive control method for core power control in pressurized water reactor nuclear power stations *Nucl. Eng. Technol.* **49** 134–140.
[20] Bemporad A, Ricker N L and Morari M 2021 *Model predictive control toolbox getting started*

guide (The MathWorks).

Acknowledgements
This work was supported by a Fundamental Research Grant Scheme (FRGS) grant (FRGS/1/2017/TK04/UTM/02/55) from the Ministry of Higher Education, Malaysia. The author would like to thank all individuals who have contributed to this paper especially Dr Nurul Adilla Binti Mohd Subha from Universiti Teknologi Malaysia and to the Malaysian Nuclear Agency under the Ministry of Science, Technology, and Innovation, Malaysia (MOSTI) for the support.

Lattice Design of a Compact Hadron Driver for Cancer Therapy

Leo Kwee Wah[a], Ken Takayama[b], Toshikazu Adachi[b], Tadamichi Kawakubo[b] and Tanuja Dixit[c]
[a]Nuclear Agency, Bangi 43000 Kajang. Selangor. Malaysia
[b]High Energy Accelerator Research Organization (KEK), Tsukuba, Ibaraki, Japan.
[c]*Society for Applied Microwave Electronics Engineering & Research (SAMEER), Mumbai, India*

Abstract.. A next generation cancer therapy driver by using a fast-cycling induction synchrotron is given. The design of the lattice parameters with zero momentum-dispersion and high-flat momentum-dispersion region as essential parameters must be fulfilled and the optimization process with results are discussed here. Finally, the optimization process and method are developed and the lattice parameters are well defined.

1. Introduction

A design of a compact hadron driver for cancer therapy based on the induction synchrotron concept is given in Refs. 1 and 2 with parameters as in Table 1 and schematics in Fig. 1. This lattice has two-fold symmetry with a circumference of 52.8 m, a 2 m-long dispersion-free straight section, and a 3 m-long large flat dispersion straight section. By assuming a 1.5 T bending magnet, the ring can deliver heavy ions of 200 MeV/au at 10 Hz.

Table 1. Compact hadron driver parameters

Parameters	Specification
Energy	200 MeV/nucleon for A/Q = 2 ion
Circumference, C_0	52.8 m
Ion Species	Gaseous/Metal ions
Ion Source (IS)	Laser ablation IS, Electron Cyclotron Resonance IS
Injector	200 kV (electrostatic)
Ring	Fast cycling (10 Hz)
	B_{max} = 1.5 Tesla
	ρ = 2.8662 m
	FODOF cell with edge focus of B
	Mirror symmetry
	ν_x/ν_y = 1.3143/1.4635
	2m long dispersion-free region
	3m long flat large dispersion region
	α_p=0.273088
	γ_T=1.92
	Induction cells driven by SPS employing SiC-MOSFET
Acceleration	$V_{acc} = \rho C_0 dB/dt$ (max 7 kV)
Vacuum	10^{-8} Pascal

In order to realize the slow extraction technique and fast extraction in a fast-cycling synchrotron, which allows the energy sweep beam scanning, the essential features [1,2] required are:

i. Dispersion-free region for induction acceleration devices and injection device
ii. Localized large flat dispersion region for the slow extraction with the length of 3 m
iii. Local betatron phase advance of π/2 for the fast extraction

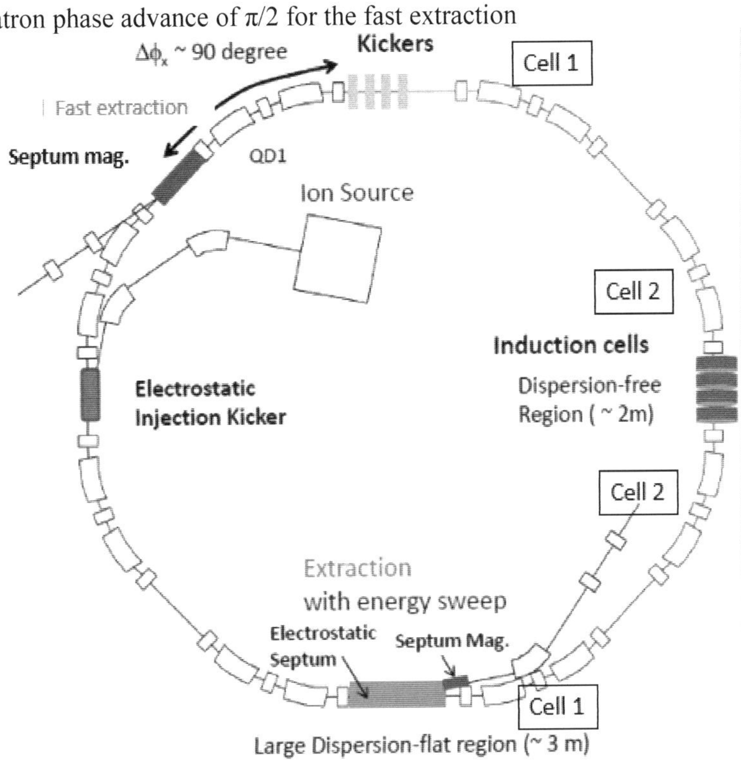

Figure 1. Hadron Driver System Schematics

Details of the lattice parameters and optimization method are discussed in the next section.

2. Lattice: Basics and Configuration

The lattice of compact hadron driver is designed based on the Proton Ion Medical Machine Study (PIMMS) [3]. A lattice is an array of bending magnets and focusing magnets. Whereby the coefficient of restoring force, K(s), in the betatron equation is uniquely determined by this lattice. Hence, the individual common components can be treated with 3 X 3 matrices with the following equations [4,5,6].

Drift space transfer matrix,

$$M_D = \begin{bmatrix} 1 & \ell & 0 \\ 0 & 1 & 0 \\ 0 & 0 & 1 \end{bmatrix} \tag{1}$$

magnet transfer matrix,

$$M_f = \begin{bmatrix} \cosh \sqrt{K_F}\, \ell_{Q_F} & \dfrac{\sinh \sqrt{K_F}\, \ell_{Q_F}}{\sqrt{K_F}} & 0 \\ \sqrt{K_F} \sinh \sqrt{K_F}\, \ell_{Q_F} & \cosh \sqrt{K_F}\, \ell_{Q_F} & 0 \\ 0 & 0 & 1 \end{bmatrix} \tag{2}$$

Defocusing magnet transfer matrix,

$$M_d = \begin{bmatrix} \cos \sqrt{K_D}\, \ell_{Q_D} & \dfrac{\sin \sqrt{K_D}\, \ell_{Q_D}}{\sqrt{K_D}} & 0 \\ -\sqrt{K_D} \sin \sqrt{K_D}\, \ell_{Q_D} & \cos \sqrt{K_D}\, \ell_{Q_D} & 0 \\ 0 & 0 & 1 \end{bmatrix} \tag{3}$$

Bending magnet transfer matrix,

$$M_B = \begin{pmatrix} \cos\theta & \rho \sin\theta & \rho(1-\cos\theta) \\ -\sin\theta/\rho & \cos\theta & \sin\theta \\ 0 & 0 & 1 \end{pmatrix} \tag{4}$$

Horizontal edge effect transfer matrix of bending magnet,

$$M_{HE} = \begin{pmatrix} 1 & 0 & 0 \\ \tan\varepsilon/\rho & 1 & 0 \\ 0 & 0 & 1 \end{pmatrix} \tag{5}$$

Vertical edge effect transfer matrix of bending magnet,

$$M_{VE} = \begin{pmatrix} 1 & 0 & 0 \\ 1/\rho \left({}^{b}/_{6\rho \cos\varepsilon} - \tan\varepsilon \right) & 1 & 0 \\ 0 & 0 & 1 \end{pmatrix} \tag{6}$$

K- value,

$$K = \frac{B'}{B\rho} \tag{7}$$

Where:

$\theta =$ Bending angle of bending magnet
$\rho =$ Curvature of of bending magnet
$\varepsilon =$ Edge angle of bending magnet
$b =$ Distance over which the fringe field drops to zero of bending magnet
$K_{D,F} =$ K – value of the related magnets
ℓ or $\ell_{Q_{D,F}} =$ Length of the related space or magnet

The lattice configuration of compact hadron driver is shown in Fig. 2 with the half ring configuration of Cell-1 X Cell-2 X Drift space-4 X Cell-2 X Cell-1 with length of 38.2 m and the elements parameters are shown in Table 2 and Table 3.

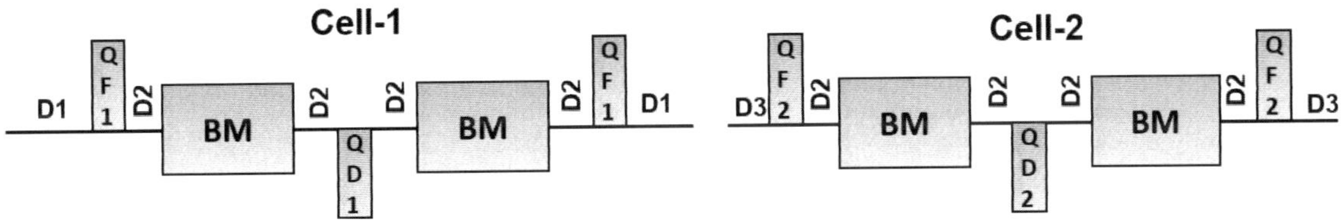

Figure 2. Configuration of Cell-1 and Cell-2

Table 2. Parameters of drift space and quadrupole elements (*After optimization)

Elements	Length, l [m]	K-value [m^{-2}]
Drift space 1, D1	1.5	---
Drift space 2, D2	0.3	---
Drift space 3, D3	0.5	---
Drift space 4, D4	1.0	---
Focusing Quadrupole magnet 1, QF1	0.3	0.75568*
Defocusing Quadrupole magnet, 1 QD1	0.3	- 0.53774*
Focusing Quadrupole magnet 2, QF22	0.3	0.40173*
Defocusing Quadrupole magnet 2, QD1	0.3	- 0.8957*

Table 3. Parameters of bending magnet

Magnetic Flux Density, B [T]	Bending Angle, θ [°]	Edge Angle, θ_E [°]	Curvature, ρ [m]	Length, l [m]
1.5	22.5	11.25	2.8662	1.1256

By referring to Eqs. (1) to (6) with parameters of Table 2 and Table 3, the transfer matrix of individual cell can be expressed as follows:

For horizontal transfer matrix of Cell 1,

$$M_{Hcell1} = M_{D1}.M_{f1}.M_{D2}.M_{HE}.M_B.M_{HE}.M_{D2}.M_{d1}.M_{D2}.M_{HE}.M_B.M_{HE}.M_{D2}.M_{f1}.M_{D1} \quad (8)$$

For horizontal transfer matrix of Cell 2,

$$M_{Hcell2} = M_{D3}.M_{f2}.M_{D2}.M_{HE}.M_B.M_{HE}.M_{D2}.M_{d2}.M_{D2}.M_{HE}.M_B.M_{HE}.M_{D2}.M_{f2}.M_{D3} \quad (9)$$

For vertical transfer matrix of Cell 1,

$$M_{Vcell1} = M_{D1}.M_{d1}.M_{D2}.M_{VE}.M_{DB}.M_{VE}.M_{D2}.M_{f1}.M_{D2}.M_{VE}.M_{DB}.M_{VE}.M_{D2}.M_{d1}.M_{D1} \quad (10)$$

For vertical transfer matrix of Cell 2,

$$M_{Vcell2} = M_{D3}.M_{d2}.M_{D2}.M_{VE}.M_{DB}.M_{VE}.M_{D2}.M_{f2}.M_{D2}.M_{VE}.M_{DB}.M_{VE}.M_{D2}.M_{d2}.M_{D3} \quad (11)$$

Whereby the total transfer matrix can be expressed as:

For horizontal transfer matrix of half ring,

$$M_{HTotal} = M_{Hcell1}.M_{Hcell2}.M_{D4}.M_{Hcell2}.M_{Hcell1} \quad (12)$$

For vertical transfer matrix of half ring,

$$M_{VTotal} = M_{Vcell1}.M_{Vcell2}.M_{D4}.M_{Vcell2}.M_{Vcell1} \quad (13)$$

For example, the horizontal transfer matrix of Cell 1 can be expressed as:

$$M_{Hcell1} = \begin{pmatrix} m_{11} & m_{12} & m_{13} \\ m_{21} & m_{22} & m_{23} \\ 0 & 0 & 1 \end{pmatrix}_{Hcell1} \tag{14}$$

$$\begin{aligned} M_{Hcell1} = & \begin{pmatrix} 1 & \ell_1 & 0 \\ 0 & 1 & 0 \\ 0 & 0 & 1 \end{pmatrix} \begin{pmatrix} \cos\sqrt{K_F}\,\ell_{Q_F} & \dfrac{\sin\sqrt{K_F}\,\ell_{Q_F}}{\sqrt{K_F}} & 0 \\ -\sqrt{K_F}\sin\sqrt{K_F}\,\ell_{Q_F} & \cos\sqrt{K_F}\,\ell_{Q_F} & 0 \\ 0 & 0 & 1 \end{pmatrix} \begin{pmatrix} 1 & \ell_2 & 0 \\ 0 & 1 & 0 \\ 0 & 0 & 1 \end{pmatrix} \begin{pmatrix} 1 & 0 & 0 \\ \dfrac{\tan\varepsilon}{\rho} & 1 & 0 \\ 0 & 0 & 1 \end{pmatrix} \\ & \times \begin{pmatrix} \cos\theta & \rho\sin\theta & \rho(1-\cos\theta) \\ -\dfrac{\sin\theta}{\rho} & \cos\theta & \sin\theta \\ 0 & 0 & 1 \end{pmatrix} \begin{pmatrix} 1 & 0 & 0 \\ \dfrac{\tan\varepsilon}{\rho} & 1 & 0 \\ 0 & 0 & 1 \end{pmatrix} \begin{pmatrix} 1 & \ell_2 & 0 \\ 0 & 1 & 0 \\ 0 & 0 & 1 \end{pmatrix} \\ & \times \begin{pmatrix} \cosh\sqrt{K_D}\,\ell_{Q_D} & \dfrac{\sinh\sqrt{K_D}\,\ell_{Q_D}}{\sqrt{K_D}} & 0 \\ \sqrt{K_D}\sinh\sqrt{K_D}\,\ell_{Q_D} & \cosh\sqrt{K_D}\,\ell_{Q_D} & 0 \\ 0 & 0 & 1 \end{pmatrix} \begin{pmatrix} 1 & \ell_2 & 0 \\ 0 & 1 & 0 \\ 0 & 0 & 1 \end{pmatrix} \begin{pmatrix} 1 & 0 & 0 \\ \dfrac{\tan\varepsilon}{\rho} & 1 & 0 \\ 0 & 0 & 1 \end{pmatrix} \\ & \times \begin{pmatrix} \cos\theta & \rho\sin\theta & \rho(1-\cos\theta) \\ -\sin\theta/\rho & \cos\theta & \sin\theta \\ 0 & 0 & 1 \end{pmatrix} \begin{pmatrix} 1 & 0 & 0 \\ \tan\varepsilon/\rho & 1 & 0 \\ 0 & 0 & 1 \end{pmatrix} \\ & \times \begin{pmatrix} 1 & \ell_2 & 0 \\ 0 & 1 & 0 \\ 0 & 0 & 1 \end{pmatrix} \begin{pmatrix} \cos\sqrt{K_F}\,\ell_{Q_F} & \dfrac{\sin\sqrt{K_F}\,\ell_{Q_F}}{\sqrt{K_F}} & 0 \\ -\sqrt{K_F}\sin\sqrt{K_F}\,\ell_{Q_F} & \cos\sqrt{K_F}\,\ell_{Q_F} & 0 \\ 0 & 0 & 1 \end{pmatrix} \begin{pmatrix} 1 & \ell_1 & 0 \\ 0 & 1 & 0 \\ 0 & 0 & 1 \end{pmatrix} \end{aligned} \tag{15}$$

Therefore, from Eqs. (1) to (15) the essential lattice parameters which are betatron tune, phase advance and dispersion function that can be expressed as follows:

For horizontal or vertical phase advance of half ring,

$$\mu_{HTotal,Vtotal} = \{\cos^{-1}((m_{11} + m_{22})/2)\}_{HTotal,VTotal} \tag{16}$$

Fractional part of horizontal or vertical tune with unknown integer part,

$$\Delta v_{HTotal,Vtotal} = \left\{\frac{\mu}{2\pi}\right\}_{HTotal,Vtotal} \tag{17}$$

For dispersion function at particular transfer location,

$$\begin{pmatrix} D \\ D' \end{pmatrix}_i = \left\{\frac{1}{2-(m_{11}+m_{22})}\begin{pmatrix} m_{13} + m_{12}m_{23} - m_{22}m_{13} \\ m_{23} + m_{21}m_{13} - m_{11}m_{23} \end{pmatrix}\right\}_i \tag{18}$$

Whereby, Eqs. (16), (17) and (18) are essential to optimize the K-value of individual quadrupole elements as described at next section.

3. Lattice: Optimization Process and Result

For such evaluation, momentum deviation is assumed as 0.01 and three cases have been evaluated as follows. The lattice of hadron driver can be simplified with four FODO cells and treated as a super-period. They are grouped into two types of cells, CELL-1 and CELL-2, and aligned as CELL-1, CELL-2, Drift4, CELL-2 and CELL-1 as shown in Fig. 3.

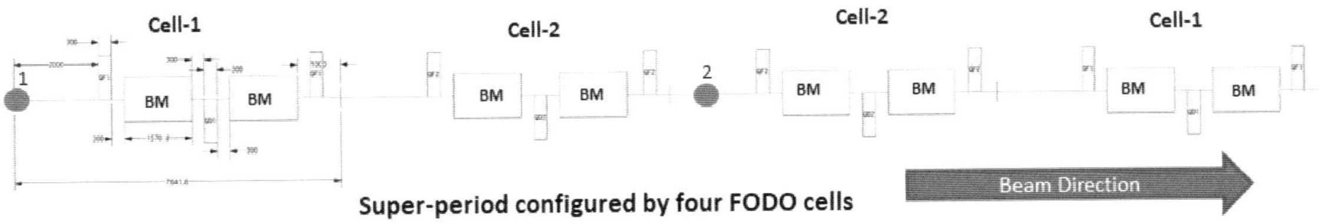

Figure 3. Configuration of half ring (1 denote as large flat dispersion region and 2 denote as dispersion-free region)

The crucial requirements of this lattice are noted in Fig. 3 with large flat dispersion and dispersion-free region. Therefore, the aim is to optimize the K-value of quadrupole to achieve the (D, D')~ (arbitrary number , 0) at location 1 and (D, D') ~ (0, 0) at location 2 (Fig. 3) by scanning the vertical tune ν_V , while ν_H is fixed (D may be slightly modified). The detail process and successive iteration are shown in Fig. 4 and Fig. 5.

Choose the arbitrary K-value of the quadrupole elements

↓

Evaluate 3 χ^2's (QF1 is scanned by an appropriate small value e, while QD1 is fixed)

$$\chi_1^2 = (\Delta\nu_{H1} - \Delta\nu_{H0})^2 + (\Delta\nu_{V1} - \Delta\nu_{V0})^2$$
$$\chi_2^2 = (\Delta\nu_{H2} - \Delta\nu_{H0})^2 + (\Delta\nu_{V2} - \Delta\nu_{V0})^2$$
$$\chi_3^2 = (\Delta\nu_{H3} - \Delta\nu_{H0})^2 + (\Delta\nu_{V3} - \Delta\nu_{V0})^2$$

Find optimized QD1 (=<QF1>) by using parabola fitting to 3 χ^2's

↓

Evaluate 3 χ^2's (QD1 is scanned by an appropriate small value e, while QF1 is fixed)

$$\chi_1^2 = (\Delta\nu_{H1} - \Delta\nu_{H0})^2 + (\Delta\nu_{V1} - \Delta\nu_{V0})^2$$
$$\chi_2^2 = (\Delta\nu_{H2} - \Delta\nu_{H0})^2 + (\Delta\nu_{V2} - \Delta\nu_{V0})^2$$
$$\chi_3^2 = (\Delta\nu_{H3} - \Delta\nu_{H0})^2 + (\Delta\nu_{V3} - \Delta\nu_{V0})^2$$

Find optimized QD1 by using parabola fitting to 3 χ^2's

↓

Test change of χ^2. If test value is larger than some small value (~0.001 or less), repeat optimization procedure by setting QF1=<QF1> , QD1=<QD1> and $\chi^2_{0,previous} = \chi^2_0$.

$$test = \frac{|\chi_0^2 - \chi_{0.previous}^2|}{\chi_{0,previous}^2}$$

Figure 4. Optimization procedure of K-value

For each stage, the dispersion function at location 1 and 2 must be evaluated according to the Eq. (18). In additional, the QD_2 is used for fine tuning to fulfill the 3 lattice requirements of the compact hadron driver. Does this mean you change the length QF1 and QD1 and calculate new v using eq 15?

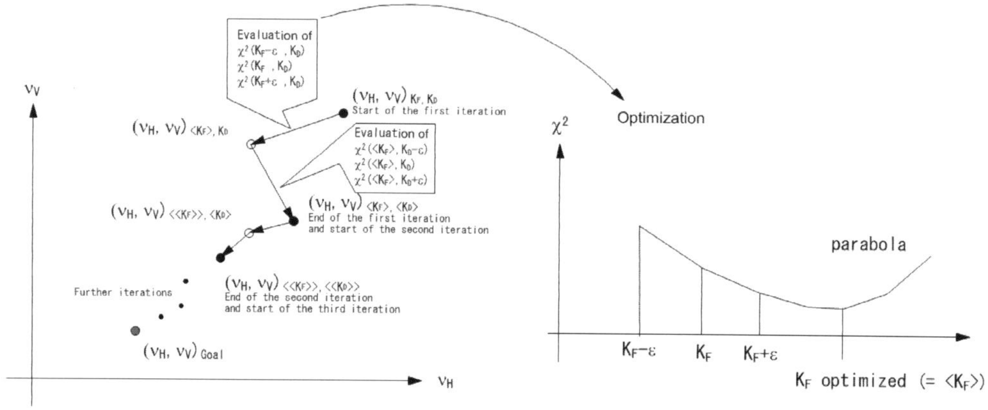

Figure 5. Schematic of tune drift during successive iteration

According to the process as described in Fig. 4, the K-value of quadrupole elements have been optimized with the result as shown in Table 4. Meanwhile, the dispersion function at location 1 and location 2 must be evaluated as well to confirm the desirable dispersion magnitude. If the large flat and zero dispersion regions haven't been realized by this optimization process. Then the optimization process should be restarted from the beginning.

Table 4. Optimization result (i, ii, iii denotes the essential lattice features of compact hadron driver)

Parameters	Specifications	Remarks
K-Value of QF_1 [m^{-2}]	0.75568	Optimized
K-Value of QD_1 [m^{-2}]	-0.53774	Optimized
K-Value of QF_2 [m^{-2}]	0.40173	Optimized
K-Value of QD_2 [m^{-2}]	-0.89570	Optimized
D_2 [m]	$-5.50X10^{-8}$	Dispersion-free region (i)
D_2' [m]	2.22×10^{-16}	Sustainable for along 2 m (i)
D_1 [m]	5.8725	Large flat dispersion region (ii)
D_1' [m]	3.24×10^{-16}	Sustainable along 3 m (ii)
Phase advance	$\pi/2$	For fast extraction (iii)

The result shown in Table 4, the lattice requirements of compact hadron driver: dispersion region, large flat dispersion region and betatron phase advance for fast extraction has been fulfilled with the optimized K-value of quadrupole elements. Hence, the beta function can be plotted by using the Courant-Snyder parameters, which is discussed in the next section.

4. Result: Courant-Snyder Parameters and Beta Function

A beam comprised of particles will move at different movements along the design orbit in a transport line, where magnets are aligned. All particles in a beam are positioned in a phase space. If such a phase space is scanned along the beam orbit, there revealed a surface enclosing all particles. This surface is called as Beam Envelope. From the standpoint of accelerator design, it is important to know the beam behavior rather than the behavior of each particle. In the phase space, the beam envelope is given by an ellipse as shown in Fig. 6, which is defined by the following formula [4,5,6]:

$$\beta x'^2 + 2\alpha x x' + \gamma x^2 = \varepsilon \tag{19}$$

Where,

β, α and γ = Courant-Snyder Parameter coefficients depending on the orbit position s.

ε = Emittance and defined by an ellipse area divided by π.

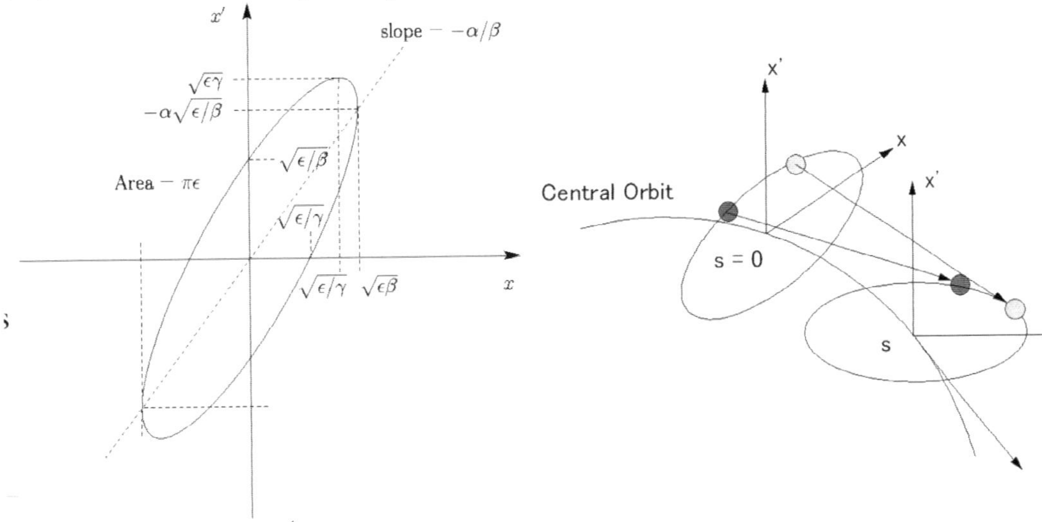

Figure 6. Courant-Snyder parameters with phase space revolution along the ring

Since the Courant-Snyder parameter satisfies the following relation, the number of independent parameters is two.

$$\beta\gamma = 1 + \alpha^2 \tag{20}$$

The Twiss parameter is transferred by a matrix constructed by a corresponding transfer matrix given as M. Here, m_{ij} is the component of the transfer matrix.

$$\begin{pmatrix} \beta \\ \alpha \\ \gamma \end{pmatrix}_s = M \begin{pmatrix} \beta_0 \\ \alpha_0 \\ \gamma_0 \end{pmatrix} \quad , \quad M \equiv \begin{pmatrix} {m_{11}}^2 & -2m_{11}m_{12} & {m_{12}}^2 \\ -m_{11}m_{21} & m_{11}m_{22} + m_{12}m_{21} & -m_{12}m_{22} \\ {m_{21}}^2 & -2m_{21}m_{22} & {m_{22}}^2 \end{pmatrix} \tag{21}$$

Whereby, the dispersion function is constructed by a corresponding transfer matrix given by M (give another subscript).. Here, m_{ij} is the component of the transfer matrix.

$$\begin{pmatrix} D \\ D' \\ 1 \end{pmatrix}_s = M \begin{pmatrix} D_0 \\ D'_0 \\ 1 \end{pmatrix}_s \quad , \quad M \equiv \begin{pmatrix} m_{11} & m_{12} & m_{13} \\ m_{21} & m_{22} & m_{23} \\ 0 & 0 & 1 \end{pmatrix} \tag{22}$$

From Eqs. (21, 22) with the lattice parameters from Table 2., the beta and dispersion functions of the compact hadron driver is plotted as shown in Fig. 7.

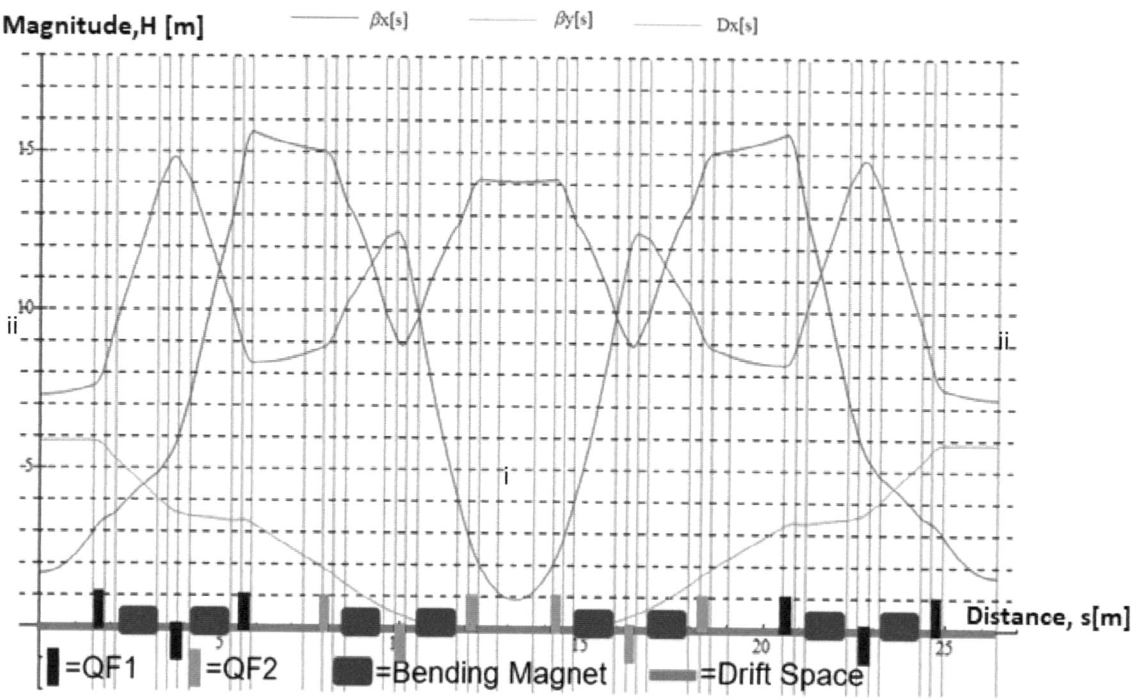

Figure 7. Beta function Vs distance of compact hadron driver

Fig. 7 shows that the requirements of the dispersion free region and the large flat dispersion region with sustainability of 4 m are fulfilled.

5. Conclusion

Using the optimization method of tune scanning, the lattice parameters of the compact hadron driver have been defined. The requirements of the dispersion free region, the large flat dispersion region and the phase advance of 90° condition for fast extraction have been fulfilled as well. For further consideration, non-linear perturbation terms must be studied and the magnetic component designs must be able to be realized with the designed lattice. Hence, the lattice should be modified to comply with the engineering aspects of the magnetic components.

6. Acknowledgments

We would like to express our gratitude to Ministry of Science, Technology and Innovation (MOSTI), Malaysia and High Energy Accelerator Research Organization (KEK), Japan for supporting the allocation to collaborate between Malaysian Nuclear Agency and KEK for this study.

7. References

[1] Peggs and Todd Satogata. Introduction to Accelerator Dynamics 1st Edition. Cambridge University Press; 1st Edition (August 7, 2017) K. W. Leo, T. Monma, T. Adachi, T. Kawakubo, T. Dixit and K. Takayama. Compact Hadron Driver for Cancer Therapies Using Continuous Energy Sweep Scanning. *Phys. Rev. ST AB* **19**, 042802 (2016).

[2] K. W. Leo, T. Adachi, T. Kawakubo, T. Monma, T. Dixit and K. Takayama. A Compact Hadron Driver for Cancer Therapies Using Continuous Energy Sweep Scanning. *Proceedings of the 13th International Conference on Heavy Ion Accelerator Technology* 2015, Yokohama.

[3] Accelerator Complex Study Group, CERN. Proton-Ion Medical Machine Study (PIMMS) Part 1. *CERN/PS 99-010 (DI)* (2 March 1999).
[4] Helmut Wiedemann. *Particle Accelerator Physics*. ISBN-13 978-3-540-49043-2 3rd ed. Springer Berlin Heidelberg New York.
[5] T. Adachi. *KEK-Nuclear Malaysia-Sokendai Accelerator School in Nuclear Malaysia.* 1st-4th December 2014
[6] Stephen Peggs and Todd Satogata. *Introduction to Accelerator Dynamics* 1st Edition. Cambridge University Press; 1st Edition (August 7, 2017)

IOP Conf. Series: Materials Science and Engineering 1231 (2022) 012003 doi:10.1088/1757-899X/1231/1/012003

Multiprocessing implementation for MCNP using Python

Muhammad Arif Sazali[1], Muhammad Syahir Sarkawi[1], Nur Syazwani Mohd Ali[1]

[1] School of Chemical and Energy Engineering, Faculty of Engineering, Universiti Teknologi Malaysia, 81310 Skudai, Johor, Malaysia

Abstract. Monte Carlo N-Particle (MCNP) is a widely-used code in nuclear engineering, but it needs high computation times due to the tracking of every single particle and interaction event. This causes shielding optimization using trial-and-error to take long times to complete. It can be solved by using multiprocessing, but this requires the MCNP source code which can be difficult to obtain. Therefore, this paper aims to suggest a solution on how Python can be used to run multiple instances of the code in shielding optimization. Two hardware setups were tested: one with a dual-core CPU, and one with a six-core CPU. Each of them would run several repeated simulations with and without multiprocessing enabled. Comparisons were made between each case of the same setup to observe the improvements in completion time. A tutorial of the Python algorithm is also provided in the methodology.

1. Introduction

Monte Carlo N-Particle (MCNP) is a nuclear simulation code developed and maintained by Los Alamos National Laboratory [1]. It is one of the nuclear modeling software that utilizes the Monte Carlo method. Other examples include Geant [2], Fluka [3], OpenMC [4], and SuperMC [5]. These codes provide very accurate results for complex situations and geometries. However, they are known to be computationally expensive, needing more calculation and time as the complexity increases. This is because the software needs to track every particle from its birth to its death or escape from the simulation space. Several works had been done to improve the runtimes of Monte Carlo simulation for radiation transport. Rydén et al. (2018) [6] utilized four graphics processing units (GPUs) to run an MC code, simulating 3.2 billion photons/s for single-photon emission computed tomography (SPECT) reconstruction. Liang et al. (2020) [7] argued that there are some limitations to using GPUs in Monte Carlo simulations, thus they set out to test the implementation of inverse transform sampling technique on CPU, hoping that it could be used to solve the GPU limitations. Nevertheless, multiprocessing is also available natively for some versions of MCNP. However, they may not be easily obtained compared to the serial version. There is a trend of using MCNP in metaheuristic algorithm optimization which demands the improvement in MCNP runtimes. This is because in these algorithms, the MCNP will be executed multiple times until the design variables were optimized.

One of the earliest studies was by Hu et al. (2008) [8] where they coupled the MCNP code to a genetic algorithm (GA) metaheuristic to optimize the design of a composite shield in a mixed radiation situation . The result was five compositions that showed comparable attenuation with lead oxide. Cai et al. (2018) [9] used the same method to optimize a multilayer shield for uranium-235 radiation. The result was six arrangements of material that proved the reliability and consistency of the metaheuristic in shield optimization. Chen et al. (2019) [10] had also implemented the same strategy in designing the shielding of a nuclear reactor in a passenger-cargo ship from a set of 15 materials. The same application was investigated by Wilson (2019) [11] but it was for bismuth-borate glasses. They were found to be comparable to limonite concrete and barite concrete for mixed radiation conditions. In 2020, two studies were conducted: one had developed six shielding designs from polyethylene, lead, and boron [12], while the other produced a 16-layer shield for a mobile nuclear reactor system from epoxy resin, lead, boron carbide, and graphene [13].

However, most studies do not reveal the method of how the repetition or multiple instances of MCNP was done. Therefore, this article aims to demonstrate a way of coupling the MCNP code with a Python-based algorithm and to investigate the improvement that can be expected when utilizing Python multiprocessing. Two different computer setups are used to observe the increase in performance if the physical CPU cores are increased. A tutorial for the Python algorithm can be found in Section 2 and Appendices.

2. Methods and materials

2.1 Hardware setups

Two Windows 10 hardware machines are used to run the Python algorithm. The first one is a laptop from 2017 which has an Intel Core i5-6200U. It is a dual-core, four-thread processor running at 2.4 GHz and it can boost up to 2.7 GHz. The second is a desktop PC that was built in 2018, equipped with the six-core, twelve-thread AMD Ryzen 5 2600X processor. Running at a base clock speed of 3.6 GHz, it has the capability of running at 4.2 GHz on all cores, provided that adequate cooling is maintained. This desktop is run using two cases: (1) all cores capped at 2.7 GHz, and (2) all cores capped at 3.9 GHz. The former is implemented to estimate the effect in the laptop setup if its logical cores were to increase beyond 4.

2.2 MCNP simulation model

Figure 1 shows the model of MCNP simulation that is run by the algorithm. Its code can be found in Appendix A. The model is a simple radiation attenuation experiment setup with a single monoenergetic source, a slab of shielding, and a detector volume. The source emits neutrons of 1-MeV in the direction of the detector (source direction biasing). The shield is made up of polyethylene and its dimension is set to 3 cm (W) × 3 cm (H) × 3 cm (L). The detector volume is a boron trifluoride-filled cylinder of 2 cm in diameter and 5 cm in length. F4 tally for neutrons is specified in the volume, while secondary photons are neglected. All 20 runs are performed at 2×10^6 nps and they are made to be independent of one another by having distinct seed numbers. A good summary of all the functions in MCNP can be found from the work of Shultis and Faw (2011) [14].

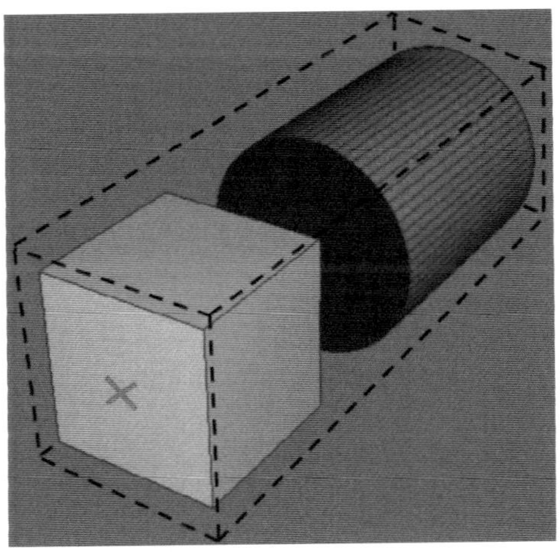

Figure 1. MCNP simulation model. The red cross represents the point source which is placed close to the surface of the polyethylene shield (yellow box). The purple cylinder is the detector volume. Both bodies are surrounded by air (dashed line black box).

IOP Conf. Series: Materials Science and Engineering 1231 (2022) 012003 doi:10.1088/1757-899X/1231/1/012003

2.3 Python multiprocessing for MCNP

The algorithm is developed using Anaconda, a distribution package of Python for scientific computing. It is user-friendly and the results of the code can be seen automatically in the same interface as the code. This means that the code can be written, run, and debugged conveniently. The algorithm is written in a Jupyter notebook (.ipynb) which is opened through the Anaconda Prompt. If done otherwise, it cannot access the MCNP cross-section file "xsdir". Figure 2 shows the flowchart of the algorithm, while the code can be found in Appendix B.

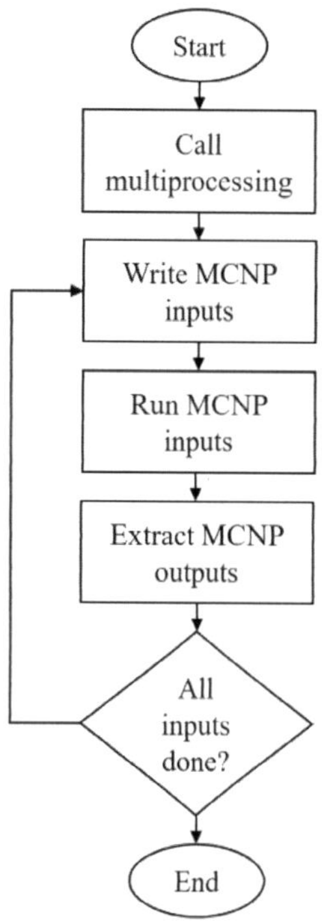

Figure 2. Python algorithm flowchart

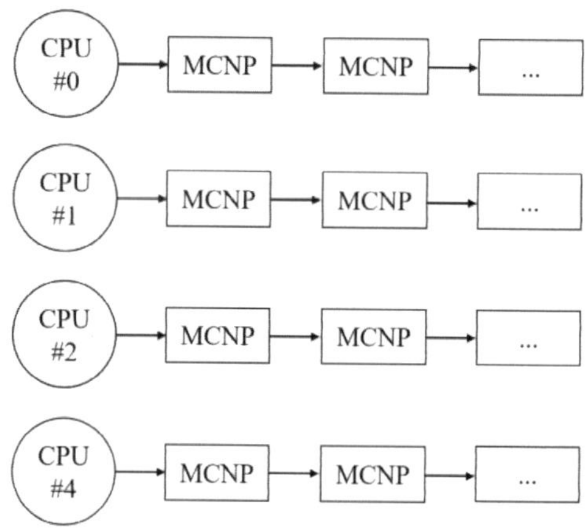

Figure 3. Illustration of the Python multiprocessing

Firstly, the multiprocessing function is called. For clarification, this function tells each logical core (thread) to run a single MCNP input at a time (Figure 3) rather than all of them run a single input. Therefore, the maximum number for the "Pool" depends on the number of logical cores. For example, the highest number for the desktop setup is 12 since it has 12 threads. After that, the "os" function is imported to enable the algorithm to run MCNP code from the command prompt. The mcnp.exe and the .ipynb file also need to be in the same folder.

Then, the MCNP input files are created using the algorithm so that each of them has a distinct seed number. This is done by defining a new function named "mcnp_in". The code for this function can be found in Appendix C. After the MCNP input lines have been added, the input is set to be run by using the command "os.system()". After the run is completed, the output file is read to extract the desired output such as tallies and relative errors. At the end of the function, all dump files ("runtpe", "runtpf", etc.) are deleted so that the next MCNP runs will not be exhausted of letters for their dump file name.

The multiple instances of MCNP runs are executed using the command "pool.starmap(mcnp_in, zip(input))". For example, if the Pool is set to 12 and there are 20 MCNP runs, the algorithm will write

12 MCNP input files and run them simultaneously. It does not wait until all 12 runs are completed to start on the remaining 8 runs. Instead, it creates and runs new ones as the current 12 runs finish. The time taken for all runs to be completed is recorded using the "time" function.

3. Results and discussion

Table 1 shows the results of the experiment for logical cores, while Figure 4 plots the completion time against the number of logical cores. Note that when there is only one logical core, it means that the MNCP runs are executed in series using one logical core. Based on Table 1, there are no significant changes to the average tally and the average relative error when the number of logical cores increases. The average relative errors for all cases are also constant at 0.0038 which is less than 0.05. Although this does not guarantee the acceptability of the tally, they are reliable based on the tally interpretation [12].

Table 1. Experimental results for the number of logical cores

No. of logical cores	Laptop (2.7 GHz)			Desktop (2.7 GHz)			Desktop (3.9 GHz)		
	Time taken (s)	Ave. tally (10^{-3} cm^{-2})	Ave. rel. error (10^{-3})	Time taken (s)	Ave. tally (10^{-3} cm^{-2})	Ave. rel. error (10^{-3})	Time taken (s)	Ave. tally (10^{-3} cm^{-2})	Ave. rel. error (10^{-3})
1	395.79	1.06282	3.8	426.26	1.06322	3.8	283.93	1.06301	3.8
2	247.34	1.06325	3.8	234.63	1.06262	3.8	159.18	1.06262	3.8
4	180.58	1.06376	3.8	128.53	1.06160	3.8	88.27	1.06164	3.8
6	[redacted]			86.94	1.06208	3.8	60.18	1.06122	3.8
8				72.98	1.06212	3.8	50.89	1.06390	3.8
10				57.27	1.06020	3.8	40.22	1.06181	3.8
12				56.67	1.06091	3.8	39.57	1.06091	3.8

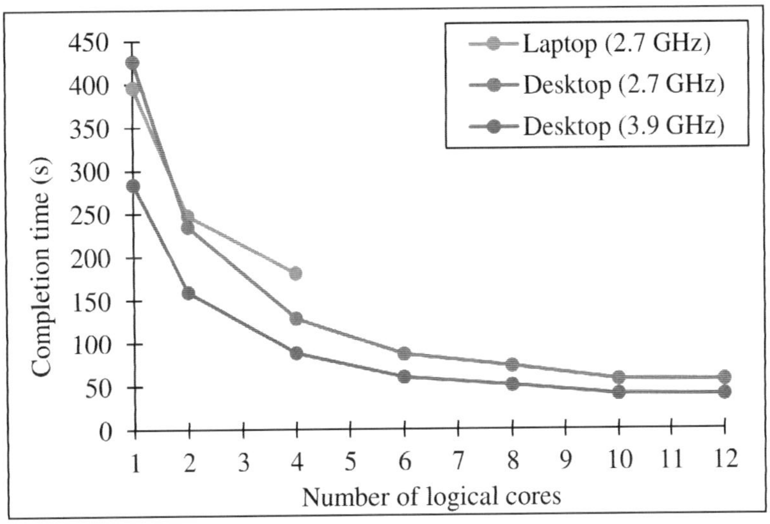

Figure 4. The completion times per number of logical cores

From Figure 4, both laptop and desktop (27 GHz) setups show a similar trend. When doubling the logical cores from 1 to 2 (or from series to parallel), the time taken for the laptop and the desktop decreases by 37.5% and 45%, respectively. The noticeable discrepancy in improvement might be due to other differences in hardware (CPU cache, RAM amount and speed, storage speed, etc.). When the logical cores are increased to 4, the desktop shows a 70% decrease in time, while the laptop shows a 54% decrease. At 6 and 8 logical cores, the time taken for the desktop decreases by 80% and 83%, respectively. The remaining 10 and 12 logical cores only decrease the MCNP runtime at very similar rates of 86.6% and 86.7%, respectively. These results suggest that the improvement per increase in logical cores is not linear and there seems to be a diminishing return after a certain number of logical cores. The reason might be because at 1 to 6 logical cores, the MCNP runs are each assigned to one thread of the 6 physical cores. Further increase in the logical cores means that the physical cores are now being shared by two threads, reducing the performance gain. Other processes that are running in the background might also cause a decrease in improvement. Nevertheless, it is clear that by having more logical cores, the MCNP runs can be finished in significantly less runtimes.

Table 2 shows the results when the CPU clock speed increases. Similar to Table 1, the average tally and relative error show no significant changes with the clock speed. Figure 5 shows the details of improvement per GHz increase on the desktop setup. It is observed that increasing the clock speed reduces the MCNP completion times. The trends suggest that their relationship is linear. However, the gradient decreases with the number of logical cores. For 1 logical core, the time taken decreases by 121.35 seconds per GHz. However, increasing the logical core by 1, the gradient decreases to 66.95 seconds per GHz (45% less than the previous case). At the maximum number of logical cores, the completion time only decreases by 17.73 seconds per GHz. This trend shows the same diminishing returns as seen in Figure 4. There might be other factors that caused these results such as the way that instructions per core (IPC) work, a limited amount of cache memory, memory latency, and other hardware or software limitations/bottlenecks.

Table 2. Experimental results for the clock speed

Clock speed (GHz)	Number of logical cores								
	1			2			12		
	Time taken (s)	Ave. tally (10^{-3} cm^{-2})	Ave. rel. error (10^{-3})	Time taken (s)	Ave. tally (10^{-3} cm^{-2})	Ave. rel. error (10^{-3})	Time taken (s)	Ave. tally (10^{-3} cm^{-2})	Ave. rel. error (10^{-3})
2.4	480.54	1.06137	3.8	264.79	1.06316	3.8	67.57	1.06091	3.8
2.7	426.26	1.06322	3.8	234.63	1.06262	3.8	56.67	1.06117	3.8
3.0	384.87	1.06382	3.8	211.62	1.06202	3.8	51.88	1.06215	3.8
3.3	349.99	1.06213	3.8	192.49	1.06336	3.8	46.81	1.06433	3.8
3.6	320.56	1.06221	3.8	176.35	1.06120	3.8	42.97	1.06404	3.8
3.9	296.10	1.06267	3.8	162.99	1.06255	3.8	39.57	1.06391	3.8

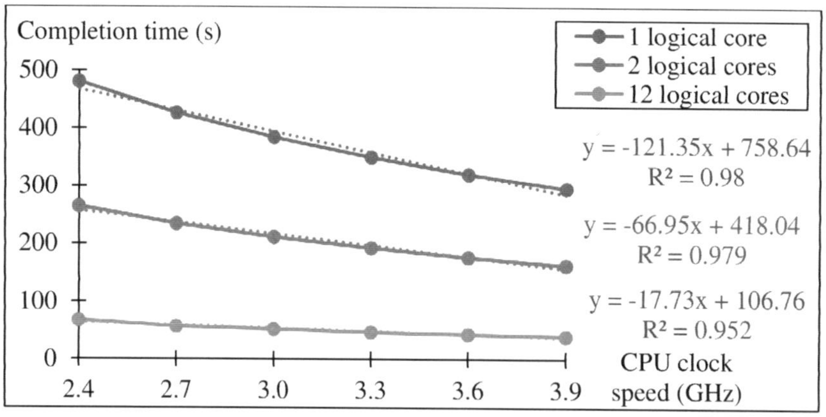

Figure 5. The completion times per CPU clock speed

4. Conclusion

A Python algorithm is developed to investigate the improvement in MCNP runtimes that can be expected when multiprocessing is enabled. 20 MCNP simulations are run using two hardware setups: a laptop with a dual-core, 4-thread processor and a desktop with a 6-core, 12-thread processor. The tutorial for writing the algorithm is also provided and all relevant codes can be found in the appendices.

By enabling multiprocessing or running MCNP in parallel, the runtimes can be decreased by at least 37.5%. Both setups show a similar trend when the number of logical cores is increased. However, the improvement per increase in logical cores is not linear and diminishing returns are observed when the logical cores exceed the physical core count. Increasing the clock speed also decreases the time taken, but the gradient reduces as the logical cores increase. Therefore, it is recommended to prioritize the number of logical cores over the CPU clock speed when running multiple instances of MCNP for metaheuristic optimizations.

Future works can be directed towards expanding the algorithm to check for the acceptability of the tally results automatically. Photons can be added into the MCNP input so that mixed radiation can be simulated which is the usual case for metaheuristic optimizations. An investigation can be done on running the Python-MCNP algorithm using discrete graphics processing units (GPU) especially Nvidia GPUs. They typically have several hundred to thousands of cores. It is interesting to observe the improvements that these GPUs can provide to parallel MCNP simulations.

5. Appendices

Appendix A. MCNP input file

```
c 3 cm PE, 1 MeV neutrons
c cell card
   1   1      -0.001205 -1 2 3 imp:n=1
   2   2      -0.93 -2 imp:n=1
   3   3      -0.002831 -3 imp:n=1
 999   0      1 imp:np=0

c surface card
   1 rpp 0 11 -2.1 2.1 -2.1 2.1   $air
   2 rpp 1 4 -1.5 1.5 -1.5 1.5    $shield
   3 rcc 6 0 0 5 0 0 2            $detector

c data card
mode n
nps 2e6
rand seed = 39360440332897
m1 6012 -0.00012 7014 -0.75527 8016 -0.23178 18000 -0.01283 $air
m2 1001 -0.1437 60000 -0.8563 $polyethylene
m3 5011 -0.15944 9019 -0.84056 $BF3
sdef pos=0.99 0 0 par=1 erg=1 vec=1 0 0 dir=d2 wgt=0.5
si2 0 1
sp2 01
f4:n 3
```

Appendix B. Python multiprocessing algorithm

```
import time
star_time = time.time()
from multiprocessing.dummy import Pool as ThreadPool
import os
import numpy as np

pool = ThreadPool(4)
run = np.arange(1,21)
tally = error = []
resultn = pool.starmap(mcnp_in,zip(run))
print (resultn)
for i in range(0,len(resultn),1):
        tally = tally + [resultn[i][1][0][0]]
        error = error + [resultn[i][1][0][1]]

print("--- %s secods ---" % (time.time() – start_time))
```

Appendix C. MCNP input function

```
def mcnp_in(x):
    rseed = int(np.random.random()*1e14)
    a = open("%sn" %(x), "a+")

    a.write('c 3 cm PE, 1 MeV neutrons \n')
    a.write('c cell card \n')
    a.write('     1       1       -0.001205 -1 2 3 imp:n=1 \n')
    a.write('     2       2       -0.93 -2 imp:n=1 \n')
    a.write('     3       3       -0.002831 -3 imp:n=1 \n')
    a.write('   999       0       1 imp:np=0 \n')
    a.write(' \n')
    a.write('c surface card \n')
    a.write('     1 rpp 0 11 -2.1 2.1 -2.1 2.1      $air \n')
    a.write('     2 rpp 1 4 -1.5 1.5 -1.5 1.5       $shield \n')
    a.write('     3 rcc 6 0 0 5 0 0 2               $detector \n')
    a.write(' \n')
    a.write('c data card \n')
    a.write('mode n \n')
    a.write('nps 2e6 \n')
    a.write('rand seed = 39360440332897 \n')
    a.write('m1 6012 -0.00012 7014 -0.75527 8016 -0.23178 18000 -0.01283 $air \n')
    a.write('m2 1001 -0.1437 60000 -0.8563 $polyethylene \n')
    a.write('m3 5011 -0.15944 9019 -0.84056 $BF3 \n')
    a.write('sdef pos=0.99 0 0 par=1 erg=1 vec=1 0 0 dir=d2 wgt=0.5 \n')
    a.write('si2 0 1 \n')
    a.write('sp2 01 \n')
    a.write('f4:n 3 \n')

    a.close()
```

```
os.system("mcnp5.exe i=%sn o=%so" %(x,x))

b.open("%so" %(x), "r")
data = b.readlines()
d = []
for i in range (0,len(data),1):
      ta1 = data[i].find("1tally   4")
      if ta1 != -1:
                  d = d + [np.fromstring(data[i+9],dtype=float,sep=' ')]
b.close()

os.system("del %sn" %(x))
os.system("del %so" %(x))
os.system("del runtpe,runtpf,runtpg,runtph,runtpi,runtpj,runtpk,runtpl")
return x,d
```

6. References

[1] Brown F B, Barrett R F, Booth T E, Bull J S, Cox L J, Forster R A, ... and Sweezy J 2002 MCNP version 5. *Trans. Am. Nucl. Soc* **87(273)** 02-3935.
[2] Brun R, Urban L, Carminati F, Giani S, Maire M, McPherson A, ... and Patrick G 1993 *GEANT: detector description and simulation tool* **CERN-W-5013**
[3] Battistoni G, Cerutti F, Fasso A, Ferrari A, Muraro S, Ranft J, ... and Sala P R 2007 The FLUKA code: Description and benchmarking *AIP Conf. proc.* **896** 31-49
[4] Romano P K and Forget B 2013 The OpenMC monte carlo particle transport code *Ann. of Nuc. Ene.* **51** 274-281
[5] Wu Y, Song J, Zheng H, Sun G, Hao L, Long P, ... and FDS Team 2015 CAD-based Monte Carlo program for integrated simulation of nuclear system SuperMC *Ann. of Nuc. Ene.* **82** 161-168
[6] Rydén T, Lagerlöf J H, Hemmingsson J, Marin I, Svensson J, Båth M, ... and Bernhardt P 2018 Fast GPU-based Monte Carlo code for SPECT/CT reconstructions generates improved ^{177}Lu images *EJNMMI phy.* **5(1)** 1-12
[7] Liang Y, Muhammad W, Hart G R, Nartowt B J, Chen Z J, James B Y, ... and Deng J 2020 A general-purpose Monte Carlo particle transport code based on inverse transform sampling for radiotherapy dose calculation *Sci. rep.* **10(1)** 1-18
[8] Hu H, Wang Q, Qin J, Wu Y, Zhang T, Xie Z, ... and Zhan Y 2008 Study on Composite Material for Shielding Mixed Neutron and γ-Rays *IEEE Tran. on Nuc. Sci.* **55(4)** 2376-2384
[9] Cai Y, Hu H, Pan Z, Hu G and Zhang T 2018 A method to optimize the shield compact and lightweight combining the structure with components together by genetic algorithm and MCNP code. *Appl. Rad. & Isot.* **139** 169-174.
[10] Chen Z, Zhang Z, Xie J, Guo Q and Yu T 2019 Metaheuristic optimization method for compact reactor radiation shielding design based on genetic algorithm *Ann. of Nuc. Ene.* **134** 318-329
[11] Wilson M 2019 Optimization of the radiation shielding capabilities of bismuth-borate glasses using the genetic algorithm *Mat. Che. and Phy.* **224** 238-245
[12] Hu G, Hu H, Yang Q, Yu B and Sun W 2020 Study on the design and experimental verification of multilayer radiation shield against mixed neutrons and γ-rays *Nuc. Eng. & Tec.* **52(1)** 178-184
[13] Hu G, Shi G, Hu H, Yang Q, Yu B and Sun W 2020 Development of gradient composite shielding material for shielding neutrons and gamma rays *Nuc. Eng. & Tec.* **52(10)** 2387-2393

[14] Shultis J K and Faw R E 2011 *An MCNP primer* 1-42

IOP Conf. Series: Materials Science and Engineering 1231(2022) 012004 doi:10.1088/1757-899X/1231/1/012004

New exposure room shielding incorporated with ferro boron concrete for neutron radiography imaging (NURI) facility at TRIGA PUSPATI Research Reactor

Muhammad Syahir Sarkawi [1, a)], Jasman Zainal [1], Muhammad Arif Sazali [1], Nur Syazwani Ali [1], Nor Afifah Basri [1], Muhammad Rawi Mohamed Zain [2], Faridah Idris [2], Khair'iah Yazid[2], Norriza Mohd Isa [2], Rafhayudi Jamro[3]

[1]Nuclear Engineering Program, Department of Energy Engineering, School of Chemical and Energy Engineering, Universiti Teknologi Malaysia, 81310 Johor Bahru, Malaysia.
[2] Malaysian Nuclear Agency, 43600 Kajang, Selangor, Malaysia.
[3]Helmholtz-Zentrum Berlin (HZB), Institut Angewandte Materialforschung (EM-IAM), Hahn Meitner Platz 1, 14109 Berlin, Germany.

[a]syahirsarkawi@utm.my

Abstract. Simulation of the new exposure room shielding for Neutron Radiography Imaging (NURI) at 1 MW PUSPATI TRIGA MARK II (RTP) nuclear research reactor in Malaysian Nuclear Agency, Malaysia has been presented in this article. Previous exposure room shielding has to be disassembled due to various reasons such as radiation streaming problem and outdated since the exposure room was constructed in 1985. The NURI facility had been undergoing major upgrading for all their instrument including beam collimator, imaging system, sample positioning, instrumentation shielding, and exposure room shielding. This paper aims to simulate the radiation dose rate mapping for the new exposure room shielding for the NURI facility using the Monte Carlo simulation code of MCNPX. The exposure room shielding design consists of several modular blocks built from ordinary concrete and Ferro boron concrete. Ferro boron concrete is used to enhance the shielding performance of the facility. Results showed that exposure room shielding containing Ferro boron concrete manage to reduce the radiation by 1150 % at the critical area compared to ordinary concrete. The average radiation dose rate around the facility during operation at 1 MW reactor power ranging from 1 μSv/hr – 10 μSv/hr.

1. Introduction

Neutron was discovered in 1932 by an English physicist, Sir James Chadwick [1]. Neutron is a particle that binds together with protons in the atomic nucleus. Neutron is neutrally charged and has a mass of about 1 amu, which is nearly the same as a proton. The special characteristic of the neutron is that they interact with the nucleus of the atom rather than its electron cloud, unlike X-ray and gamma-ray [2]. Hence, the interaction force between neutrons and nuclei is not correlated with the atomic number of the element but instead depends on the isotope of the element [3]. The first photograph capture by neutron source was obtained 3 years after the discovery of neutron itself by Kallman and Kuhn in Germany [4]. Since then, the development of neutron imaging is ongoing showing that neutron is capable of many imaging applications such as in the study of bulk materials including fuel cell, cultural heritage, concrete analysis, biological samples, etc. [5-8]. With the advancement of digital technology in the last decades, neutron radiography has been shifted from a film-based inspection method for NDT towards a powerful research tool with digital imaging detection capability [9]. To this date, there are about 50 neutron radiography facilities around the world [10].

In Malaysia, the one and only nuclear research reactor is located at the Malaysian Nuclear Agency (MNA) known as the PUSPATI TRIGA MARK II (RTP). It is a pool-type reactor capable of producing 1MW power throughout the steady-state. This research reactor has been utilized for neutron irradiations for many applications and neutron radiography [11]. The main applications are neutron activation analysis (NAA), radioisotope production, small-angle neutron scattering, and neutron radiography [12]. The layout of the beam port for the RTP reactor is shown in Figure 1.

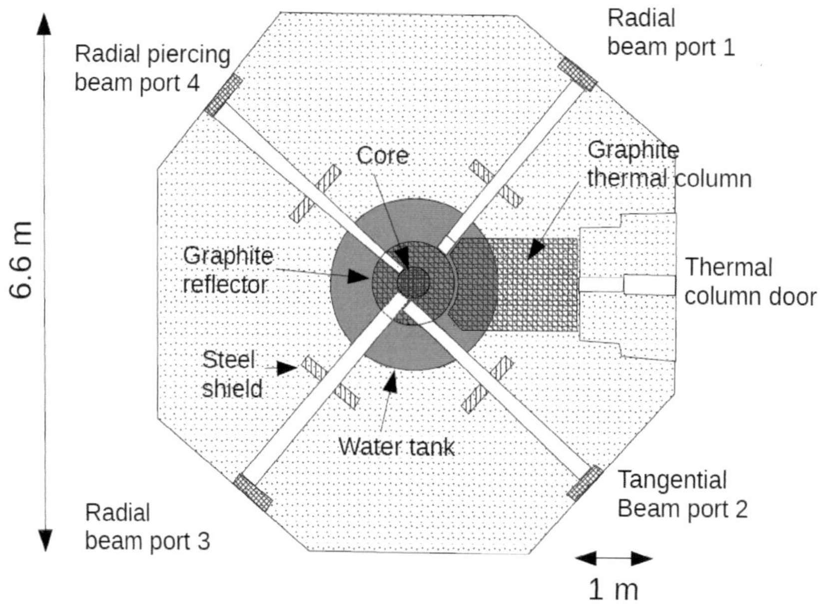

Figure 1. Top View of PUSPATI TRIGA MARK II Nuclear Research Reactor located in Malaysian Nuclear Agency [13].

The development phase of the neutron radiography facility in Malaysia has undergone three stages [11]. The first plan to construct the neutron radiography facility at RTP started soon after the research reactor was commissioned in June 1982. The first station is known as Nur-I was built from small modular concrete blocks and had been installed at radial beam port #1 [14]. The data collected from this test facility enabled the design and construction of the permanent neutron radiography facility, NUR-2. The second stage of facility development involves the relocation of instrumentations to beam port #3 and the construction of permanent exposure room shielding in 1985 as shown in Figure 2. NUR-2 had been utilized by many academia and research institutions since its operation. However, there are several limitations for NUR-2 such as low-quality neutron beam, high gamma radiation, and low collimation ratio [11, 15]. In addition, NUR-2 is only limited to the conventional film-based method due to high gamma radiation at the sample position area. Hence, due to this drawback old collimator in NUR-2 had been removed in 2014 for a new and improved collimator [15].

In the third development stage, the neutron radiography facility at RTP is moving toward digitalization same as other facilities in the world. In order to achieve this goal, a major instrumentation upgrade needs to be done. Series of studies have been done by several researchers to collect important data for designing new instrumentation such as collimator, imaging system, exposure room shielding, instruments shielding, etc. [11, 15-19]. Apart from the imaging system, exposure room shielding is one of the important components for a neutron radiography facility to ensure the safety of workers and the surrounding environment from harmful radiation exposure. In this study, a new design of exposure room shielding for the Neutron Radiography Imaging (NURI) facility at RTP is presented. This new shielding contains Ferro boron as coarse aggregates to enhance its shielding capability [20]. Initial study of using concrete containing Ferro boron had been done before, which lead to determine the ideal thickness and Ferro boron contents for the concrete [16]. The overall performance of the exposure room shielding was analyzed by using the Monte Carlo simulation codes of MCNPX.

Figure 2. Permanent Neutron Radiography Facility, NUR-2 in 1985 [14].

2. Methodology

In this work, the Monte Carlo simulation code of MCNPX version 2.6 was used to simulate the radiation dose around the new exposure room shielding for NURI. Both neutron and gamma radiation were considered in this simulation. The performance of the new exposure room shielding was tested based on the worst-case scenario in which the reactor operating at the maximum power of 1 MW. However, in an actual situation, the reactor can only be operated at 750 kW reactor power. Therefore, some fractions of the radiation might be lower than actual values. The results are shown for the total dose rate of neutron and gamma. There were 4 cases simulated in this study which are listed as follows, (Case 1) without any shielding, (Case 2) exposure room shielding with ordinary concrete only, (Case 4) exposure room shielding with ordinary and Ferro boron concrete, and (Case 5) exposure room shielding with ordinary and Ferro boron concrete including beam shutter. The beam shutter in this simulation consists of 30% borated polyethylene with a thickness of 10 cm, lead with a thickness of 15 cm, and barite colemanite concrete with a thickness of 25 cm.

The beam source spectrum and fluence at 1 MW power level were set up based on the previous study and technical report [12]. The doses were calculated using MCNPX mesh tally, with flux-to-dose conversion factors, with data sets ICRP 21-1971. This simulation is important to forecast the radiation doses around the new exposure room shielding for NURI. The input files used in this simulation were prepared as close as possible to the actual radiation source, geometry, and materials which are shown in Figure 3.

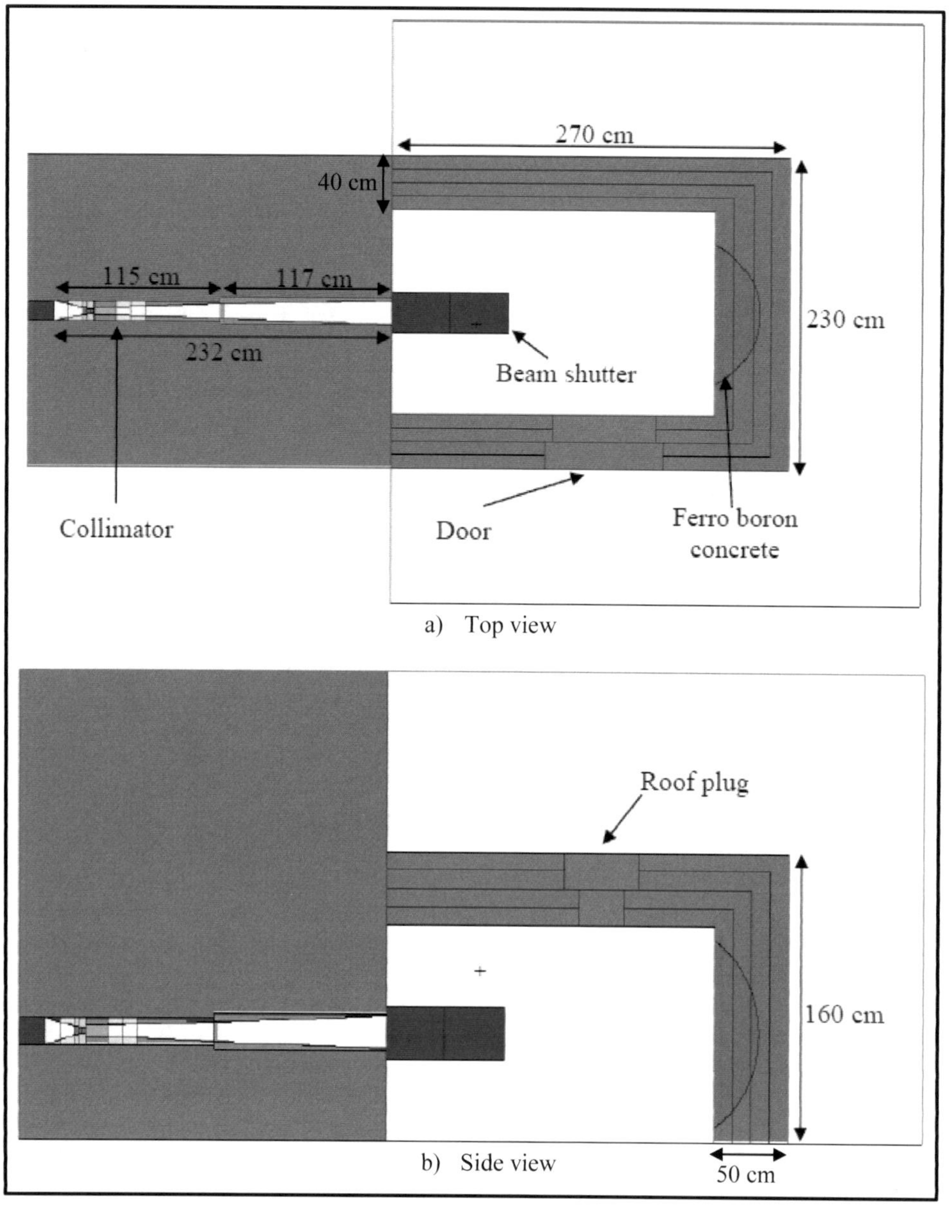

Figure 3. Side view of simulated geometry for new exposure room shielding for NURI.

3. Results and discussion

The radiation doses around the new exposure room shielding for NURI were investigated using MCNPX. The simulated results from these simulations give an estimation of the total dose rate at the NURI facility based on several cases. The simulated result for case 1 is shown in Figure 4, where a direct radiation beam was coming out from the beam port opening without any exposure room shielding. The highest dose rate measured was at the center of the beamline, with a value of approximately 0.136 Sv/hr. This showed that the most critical part that need to be addressed was the direct beam out from the beam port opening.

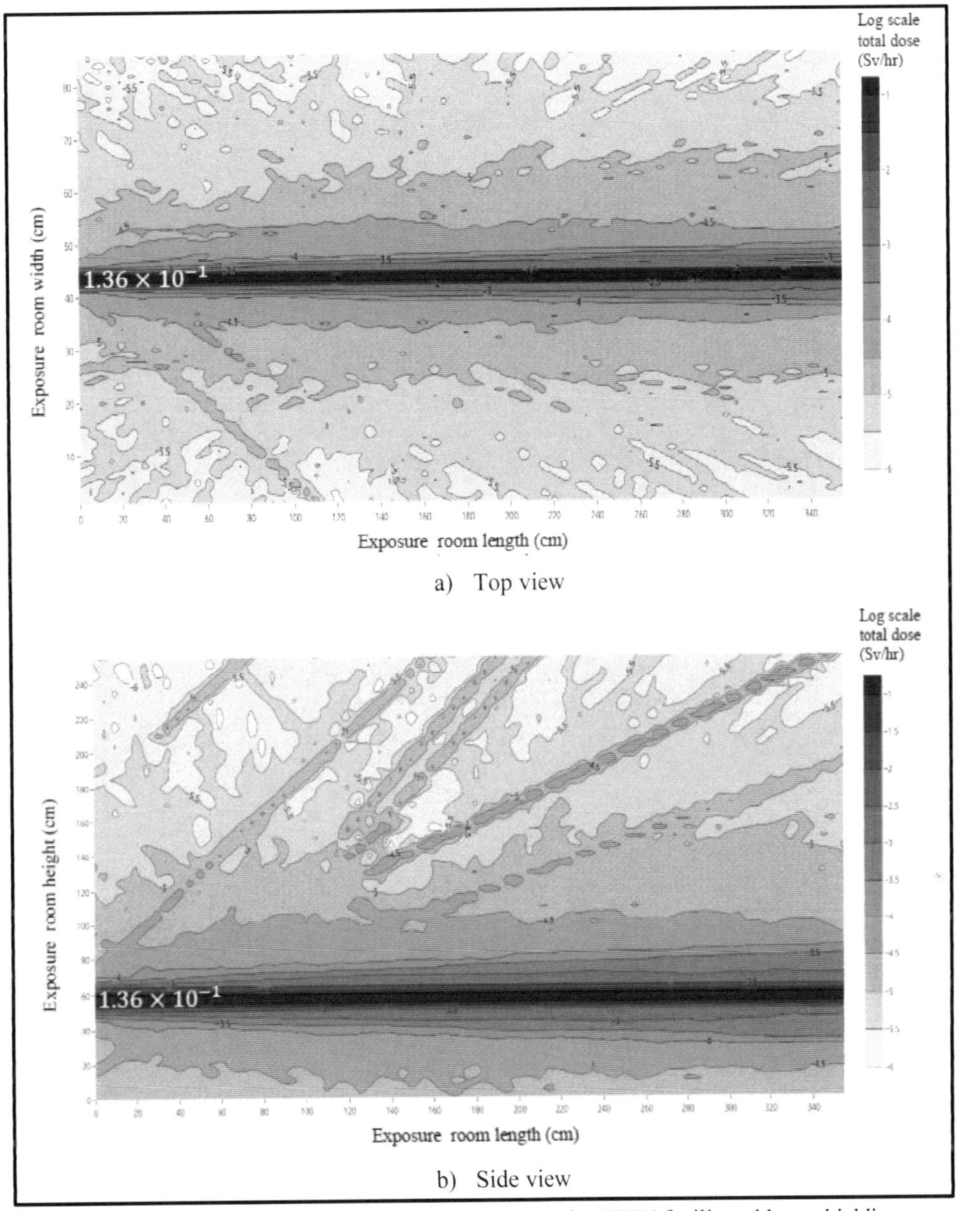

Figure 4. (Case 1) Contour plot of total dose rate at the NURI facility without shielding.

Next, case 2 was simulated with ordinary concrete only for the exposure room shielding as shown in Figure 5. The simulated results showed that the exposure room shielding managed to contain the radiation inside the exposure room. The highest rate of the total radiation dose at the outer part of the exposure room shielding was measured at the back of the shielding block surface (mark X) with a value of 3470 μSv/hr. As the value is significantly high, Ferro boron concrete was introduced in the design of the new exposure room shielding for NURI in order to reduce the radiation dose rate at that area. The other outer parts of the exposure room shielding have values ranging from 1 μSv/hr – 10 μSv/hr.

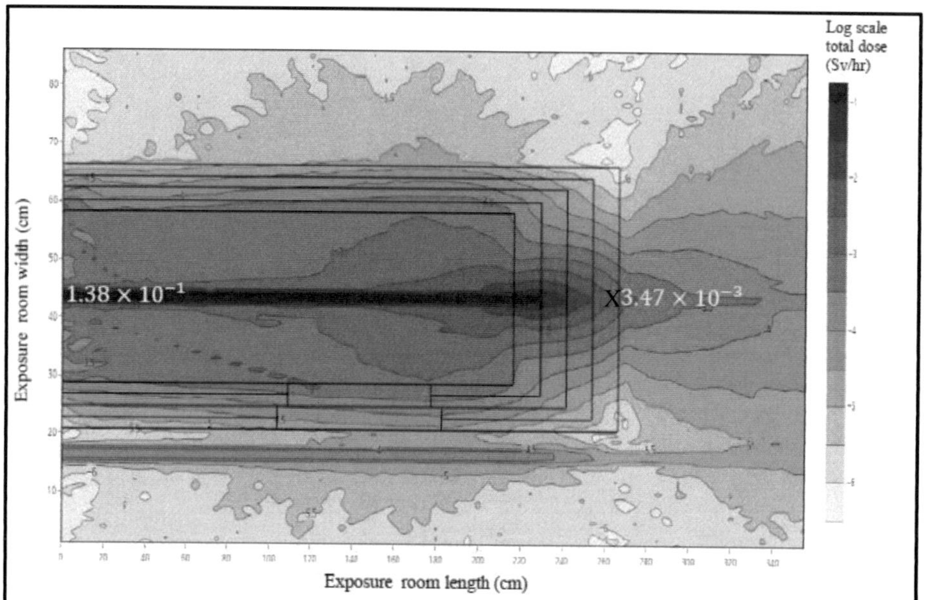

a) Top view

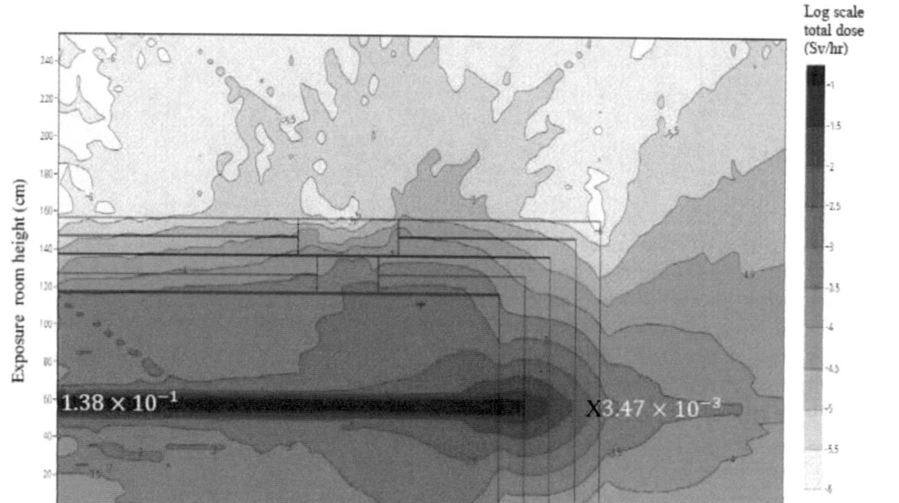

b) Side view

Figure 5. (Case 2) Contour plot of the total dose rate at NURI facility with ordinary concrete only.

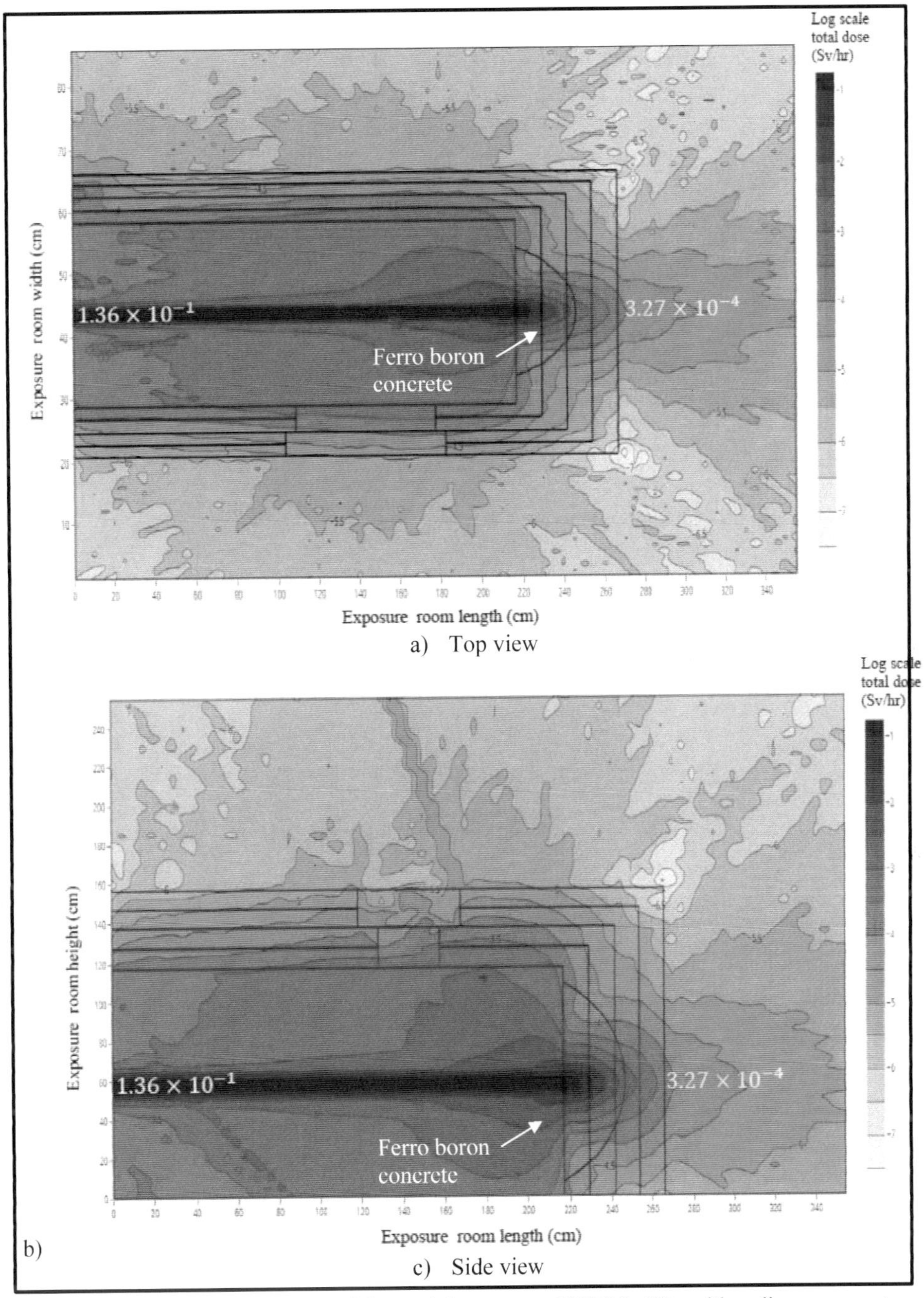

Figure 6. (Case 3) Contour plot of the total dose rate at NURI facility with ordinary concrete and ferro boron concrete

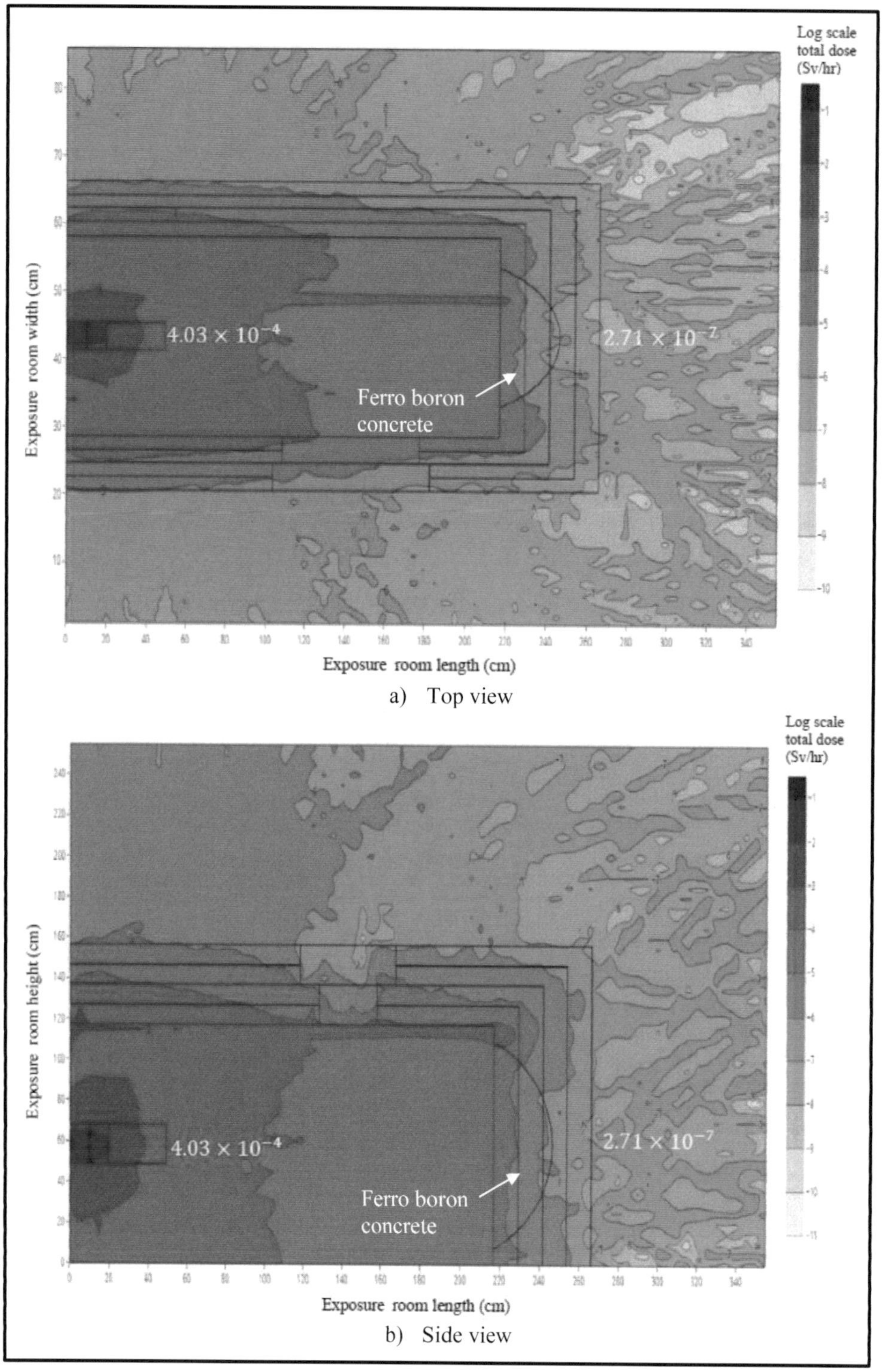

Figure 7. (Case 4) Contour plot of total dose rate at NURI facility with ordinary and Ferro boron concrete while beam shutter closed.

For case 3, the exposure room shielding was added with Ferro boron concrete at the direct beamline area. The contour plot of the total dose rate for the new exposure room shielding design at NURI is shown in Figure 6. Due to limited Ferro boron aggregates, smart approached need to be implemented in order to ensure effective shielding performance. That is why Ferro boron concrete was only be placed at a specific area not for the whole shielding blocks. From this simulated result, the dose rate at the critical area (mark x) which had been mentioned in the previous case was able to be reduced by a significant amount of about 1150%. The radiation dose rate managed to be reduced significantly due to Ferro boron concrete that has a high specific density of about 4 g/cm^3 and contains boron compounds. High-density materials are good to attenuate gamma radiation and thermalized fast neutrons. Thermalized neutron then will be captured by boron compounds in the concrete which ordinary concrete does not have. The dose rate at the back of the shielding wall surface was measured at 327 µSv/hr and the other outer parts of the exposure room shielding have values ranging from 1 µSv/hr – 10 µSv/hr. Hence, it clearly showed that Ferro boron is a good material for enhancing radiation shielding performance, especially for neutron radiation.

Case 4 was designed to simulate the total dose rate at the new NURI facility with the beam shutter closed as shown in Figure 7. Based on the simulated results, the total dose rate at the back of the beam shutter surface was 403 µSv/hr. The dose rate was reduced by a factor of 1 at 120 cm from the reactor wall. During the operation, the operator needs to consider lowering the reactor power or limiting the time when entering the exposure room even if the beam shutter is closed. The total dose rate at the outer parts of the NURI facility when the shutter closed was less than 1 µSv/hr.

Based on all the simulated cases, it shows that the new exposure room shielding at the NURI facility managed to improve its shielding capability by adding Ferro boron concrete at the wall facing directly the beamline. However, during the operation of the facility, safety measures and precaution must be taken at all time because certain areas, especially inside the exposure room has high dose rate measurement even when the beam shutter is closed. Proper procedure as stated by Atomic Energy Licencing Board (AELB) must be followed at all times during the operation [21].

4. Conclusion

The total radiation dose rate mapping for the new exposure room shielding for the NURI facility at RTP has been simulated in this study using the Monte Carlo simulation code of MCNPX. 4 cases have been simulated which the first case is without any exposure room shielding and it is to show the neutron beam at 1 MW reactor power. The second case is exposure room shielding with ordinary concrete only which is used as the control to observe the effect of the Ferro concrete. The third case is set to be as close as possible to the actual design of the exposure room shielding at NURI. The design consists of ordinary and Ferro boron concrete. The Ferro boron concrete was designed to be moulded in a semi-spherical shape in order to maximize the usage of the limited raw material of Ferro boron. Based on the simulated results, it shows that the exposure room shielding containing Ferro boron concrete manage to reduce the radiation dose rate by 1150%. It is known that Ferro boron is a great material for radiation shielding especially for neutron radiation due to boron contents [16, 20]. With the new exposure room shielding, the NURI facility at RTP can ensure the safety of all the users and the surrounding environment at all times.

5. References

[1] Berger H 1976 *Practical applications of neutron radiography and gaging* (ASTM International).
[2] Banhart J, *et al.* 2010 X-Ray and neutron imaging – complementary techniques for materials science and engineering. *Int. J. Mater. Res.*. **101** 1069.
[3] Anderson I S, McGreevy R L, and Bilheux H Z 2009 *Neutron imaging and applications : a reference for the imaging community* (New York; London: Springer).
[4] Kallmann H 1940 Photographic detection of slowly moving neutrons *US Patent No 2186757.*

[5] Zheng S *et al.* 2017 Non-destructive investigation of aluminum alloy hemmed joints using neutron radiography and X-ray computed tomography. *NDT & E Int.* **91**32-35.
[6] Zhang P, *et al.* 2017 Application of neutron radiography in observing and quantifying the time-dependent moisture distributions in multi-cracked cement-based composites. *Cement and Concrete Composites* **78** 13-20.
[7] Craft A E and Barton J P 2017 *Applications of neutron radiography for the nuclear power industry*. Phys. Procedia **88** 73-80.
[8] Fijał-Kirejczyk I M *et al.* 2013 *neutron radiography applications in studies of drying of capillary-porous systems. Drying Technol.* **31**(8) 872-880.
[9] Lehmann E H 2017 *Neutron imaging facilities in a global context.* Journal of Imaging **3**(4) 52.
[10] ISNR and IAEA 2019 Neutron imaging facilities survey, in international survey on neutron imaging facilities world-wide.
[11] Zin M R *et al.* 2020 The development of neutron imaging instrument at PUSPATI TRIGA MARK II nuclear research reactor.
[12] Rabir M H, Abdul Karim J, and Mohamed Zin M R 2017 Neutron flux characterisation of irradiation facilities in RTP *Jurnal Sains Nuklear Malaysia* **29**(2) 33-43.
[13] Farhi E *et al.* 2015 Neutron scattering instrumentation at low power reactors for science, engineering and education J. of Neutron Res., **18**(2-3) 61-77.
[14] Ramli A G *et al.* 1987 Status of Neutron Radiography in Malaysia, in Neutron Radiography (Springer) 69-76.
[15] Jamro R *et al.* 2017 Monte Carlo Simulation for Designing Collimator of the Neutron Radiography Facility in Malaysia. *Phys. Procedia* **88**(Supplement C) 361-368.
[16] Muhammad Syahir S *et al.* 2018 Radiation shielding properties of ferro-boron concrete *IOP Conf. Ser.: Mater. Sci. and Eng.* **298**(1) 012037.
[17] Sarkawi M S *et al.* 2019 Beam characterization of new neutron radiography facility at TRIGA Mark II PUSPATI research reactor. *IOP Conf. Ser.: Mater. Sci. and Eng.* **555**: 012018.
[18] Zin M R *et al.* The development of neutron imaging instrument at PUSPATI TRIGA MARK II nuclear research reactor.
[19] Yazid K *et al.* 2020Performance of imaging system at reactor TRIGA Mark II PUSPATI . *IOP Conf. Ser.: Mater. Sci. and Eng.* (IOP Publishing).
[20] Sarıyer D, Küçer R, Küçer, N 2015 Neutron shielding properties of concrete and ferro-boron. *Acta Physica Polonica A.* **128**(2B).
[21] AELB 2010 *Panduan Pengelasan Kawasan Kerja Menurut Peraturan-peraturan Pelesenan Tenaga Atom (Perlindungan Sinaran Keselamatan Asas) [P.U. (A) 46].*

Acknowledgments
The simulation software used in this paper was performed at the Malaysian Nuclear Agency, Bangi, Malaysia. The author would like to acknowledge all the contributors in completing this study.

IOP Conf. Series: Materials Science and Engineering 1231 (2022) 012005 doi:10.1088/1757-899X/1231/1/012005

Numerical solution of deuteron potential depth using python

Adam Muhammad[1], I.A. Alnour[1, (a)]

[1]Department of Physics, Faculty of Pure and Applied Sciences, International University of Africa (IUA), 12223 Khartoum, Sudan

[a]ibrahim.elnour@iua.edu.sd, ibrahim.elnour@yahoo.com

Abstract. Deuteron potential depth is hardly solvable analytically, thus this study uses a numerical method through Python programming language to calculate the depth of the potential well by coding the transcendental equation. Two different approaches were used, in which the first approach is by changing the V_{in} (initial value) value in the code to obtain the V_o values as an output, and the second, by keeping V_{in} value constant and changing the radius R values to obtain different V_o values in order to find the relationship between V_o and radius R. The mean average value of V_o was found to be –37.6 MeV. The result was close to the typical potential value (–36 MeV). The relationship between the potential depth and its radius shows the potential changes with respect to the deuteron radius. Moreover, the comparison of the deuteron potential depth of the present study shows an acceptable level of consistency with those of previous studies. Among other numerical methods, Python-based solution seems to be one of the simplest and most accurate method for solving complex numerical problems.

1. Introduction

Theoretical nuclear physics is one of the most interesting and important field that scientists have been putting much effort to understand the principle of nuclear structure and its nucleons interaction [1] and in order to achieve so scientists have tried to understand the simple structure of deuteron i.e. an isotope of hydrogen consisting of one proton and one neutron, due to its simplicity as a loosely bound system of nucleon-nucleon interaction and by understanding it, many properties of the multi-nucleon system will be understood [1, 2, 3]. For nuclear physicists, a deuteron is analogous to a hydrogen atom of atomic physicists. Just like how the measured Balmer series of electromagnetic transitions between the excited states of hydrogen led to an understanding of hydrogen structure, so should the electromagnetic transitions between the excited states of the deuteron should lead to an understanding of its structure. Unfortunately, there are no excited states of the deuteron; it is such a weakly bound system that the only "excited state" is the unbound system consisting of a free proton and a free neutron [1, 4, 5, 6, 7]. The deuteron is the only stable bound system of two nucleons. Experimentally it is known that the deuteron binding energy is 2.225 MeV, as it is known that the typical binding energy per nucleon is around 8 MeV. The deuteron is therefore a very weak bound system when (1.1 MeV) compared to a typical nucleon. It has a total spin quantum number of one and an orbital angular momentum of zero. The deuteron has a root-mean-square electromagnetic radius of approximately 2.1fm [1, 4, 8, 9].

To study the properties of deuteron based on its binding energy as a weekly bound system, we will assume that we can represent the nucleon-nucleon potential as a three-dimensional square well (Spherical well) [1].

In this study, we will solve for deuteron potential depth as one of its properties necessary to be known in order to understand the nucleon-nucleon force, but we will take a different approach from what was attempted by many researchers which solved it analytically, graphically and computationally. We will use numerical method using Python programming language to solve the transcendental equation[1] as follows:

$$K \cot KR = -L \tag{1}$$

It was well established fact that the equation was hardly solvable analytically, but numerical method was proposed by Krane in 1982 [10] and some other studies, in order to solve the problem. Computational methods via Python programming has a high reputation of solving difficult and complex physics problems and it serves as the bridge between theoretical and experimental physics.

2. Calculation

2.1. Calculation of the Potential Energy (V_o) of Deuteron

In the investigation of the deuteron potential almost all introductory textbooks represent the deuteron potential by a square potential well of depth V and width R as shown in Figure1.

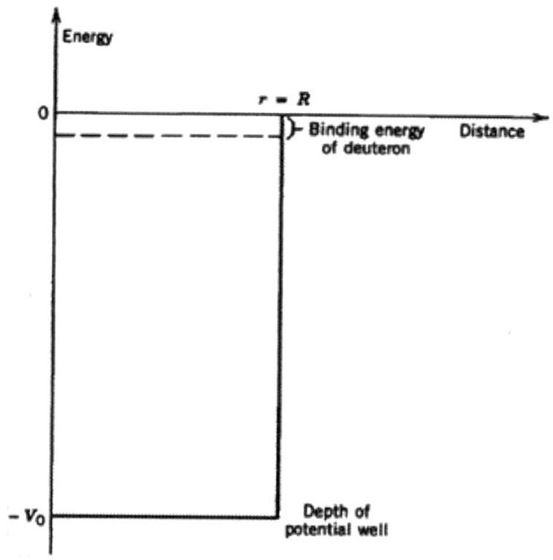

Figure1. The spherical square-potential well.

The potential of deuteron can be given by writingthe time-independent Schrödinger equation for any central potential (V_r) [11] as:

$$\hat{H}\psi(\vec{r}) = \hat{E}\psi(\vec{r})$$
$$\left[-\frac{\hbar^2}{2m}\nabla^2 + V(\vec{r})\right]\psi(\vec{r}) = \hat{E}\psi(\vec{r}) \tag{2}$$

where $V(r)$ is a central potential, ψ is the wave function, E the fundamental state binding energy, $\hbar$ is the Planck constant and m is the reduced mass which is given by:

$$m = \frac{m_p m_n}{m_p + m_n} \tag{3}$$

where m_p and m_n are masses of proton and neutron respectively.

The derivation of equation (2) was made in study [1] which eventually lead to the transcendental equation below after applying the boundary conditions to Eq. (1) above, where:

$$K = \sqrt{\frac{2m(E+V_o)}{\hbar^2}}, \quad L = \sqrt{\frac{-2mE}{\hbar^2}}$$

$$\sqrt{\frac{2m(E+V_0)}{\hbar^2}} \cot\left(R\sqrt{\frac{2m(E+V_0)}{\hbar^2}}\right) = -\sqrt{\frac{-2mE}{\hbar^2}} \quad (4)$$

The above equation (4) was then rearrange below in equation (5) to be in the form of $V_o = f(V_{in})$ in order to fit it into Python syntax

$$V_o = \tan\left(R\frac{\sqrt{2m(E+V_{in})}}{\hbar^2}\right)^2 \times E - E \quad (5)$$

2.2. Numerical Solution of Deuteron Potential through Python

Equation (5) will be solved via coding it into Python syntax, after equation (4) has been rearranged to be in the form of $X=f(x)$ i.e. $V_o = f(V_{in})$, then the algorithm is by guessing an initial V_{in} value in order to obtain the unknown V_o values, thus Python compiler iterated 10 times and it print out 10 different values among which we obtain the potential depth value (V_o) that's closest to the typical value. The same procedure was followed to find the relationship between Deuteron distance (R) and potential depth (V_o) but this time by changing the input values of R and keeping V_{in} constant, the values of V_o will be printed out and the graph of potential depth V_o versus R will be plotted from this result.

3. Results and Discussion

3.1. Calculation results of V_o

Quite reasonable result of deuteron potential depth (V_o) has been achieved from the written program. Values of V_o for different V_{in} values ranging from 1 to 1000 are shown in Table 1. The average of V_o is –37.6 MeV. The magnitude of this value shows how deep the potential well is. It is assumed that the potential depth of the deuteron only depends on the distance r between the nucleons. To be more specific it assumes that the potential has some constant value $-V_o$ (which was the one we found in this study) up to some spacing $r = R$. And it also assumes that the potential is zero for spacing larger than R.

Table 1. The results value of deuteron potential (V_o).

V_{in}	V_o(MeV)
1.0	–36.8467
1.5	–38.8395
10	–22.7267
16	–30.9610
17	–35.5801
24	–39.1824
26	–59.8322
30	–38.5322
1000	–36.0909

3.2. Deuteron potential depth as a function of R

A graph was plotted from Table 2 to show the relationship between V_o and the radius R and it was made by choosing a large number i.e. 1000 as a constant V_{in} value and changing the range of R from 0.5 up to 4 and then printing out V_o values.

Figure 2 shows depth of this potential as a function of the distance between the proton and the neutron. The curve of the graph has a minimum at $R = 1.69$ *fm*. The depth of this potential implies the interaction strength of proton and neutron and the range of this interaction is that of radius R which is 2.1 *fm* and this shows how the strong force behaves between the nucleons in the potential well as the force has a repulsive and attractive part. The nuclear force started to vanishes and become negligible at some typical spacing between the nucleons where the typical range of nuclear force is roughly 1.5 *fm* and the potential cut off R is at 2.1*fm* (In the so-called model OPEP potential)[12]. This result is in agreement with the fact that the strong force is powerfully attractive at the range of 0.5 *fm* but it decrease rapidly at the distance beyond 2.5 *fm* [1], at the distance less than 0.7 *fm* it becomes repulsive i.e. the closer the nucleons of the system are, the lower the potential energy, and the lower the potential energy, the deeper the well, and the deeper the well is, the stronger the interaction between the nucleons of that system. The fact that the potential vanishes at large nucleon spacing also appear in the wave function, which also implies that the rate of decay of the wave function at large nuclear spacing was correctly presented and this rate of decay depend only on the binding energy which is very weak. The behaviour of nuclear force became repulsive when the distance r is less than 0.5 *fm* then the potential is taken to be infinite, rather than $-V_o$. That may prevent the nucleons from getting closer to each other [1].

Table 2. Relationship between deuteron nucleon distance (R) and potential depth (V_o).

R(*fm*)	**V_o(MeV)**
0.5	–20
1.0	–41
1.5	–82
2.0	–78
2.5	–40
3.0	–54
3.5	–60
4.0	–60

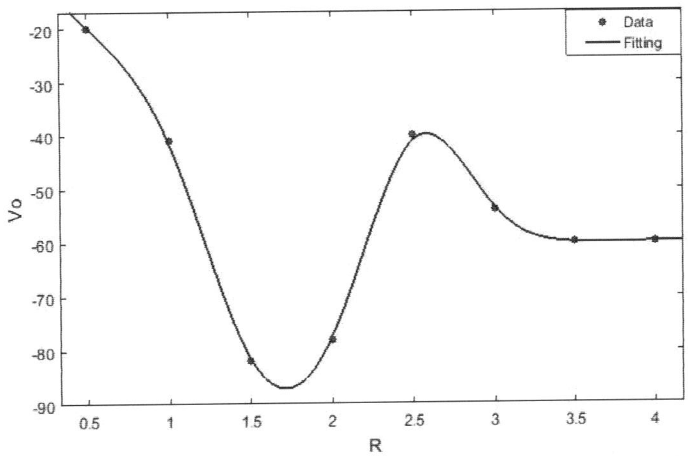

Figure 2. The $V_o$versus R Relationship for potentials having the deuteron as the ground state

3.3. Comparison with other results

A comparison between various studies [1, 12, 13, 14, 15, 16, 17] and our work is shown in Table 3, where the other studies used different methods and they arrived at different values which are quite close to each other. A lot of methods have been applied in our work, but the one that was successful was the numerical method (using python). Analytical solution was approached, and reasonably close values were achieved. Based on numerical method, the result achieved through our Python syntax was–37.6 MeV which is quite close to the typical potential value. Apparently, all the values in Table 3 are within the same range.

Finally, according to the above discussion, the potential depth value V_o obtained from Python have up to 15 digits in value (i.e. at V_{in} equals to 1; V_o = –36.846892656974795 MeV), which implies that the calculations via Python language seems to be reliable.

Table 3. The comparison between our study and other achieved values from other studies and methods

Work of Researcher	Method	V_o (MeV)
[1]	Numerical method	–35.0
[12]	A truncated one- pion exchange potential	–30.0
[13]	Modelling	–33.7
[14]	A graphical and computer method	–33.4
[15]	Analytical method	–23.1
[16]	Analytical method	–59.7
[17]	Using uncertainty principle	−51.1
Present Study	Numerical method using Python	–37.6

3.4. Estimation of the constant value for K and L based on V_o

Constants *K* and *L* were calculated based on the calculated value of V_o. *K* is found to be 0.93 $fm^{-1}$and *L* is found to be 0.23 fm^{-1}. The value of *r* was 1.69 *fm* which is estimated based on the value of *K*. The deuteron wave function is shown in Figure. 3. We can see that from Figure 3, at most of the time the wave function is outside the well and it goes to the large distance which is the implication of 37% value of the exponential. The weak binding means that ***u(r)***(wave function) is just barely able to "turn over" in the well implying that at 93% the curve started to turn down so as to connect at $r = R$ with the negative slope of the decaying exponential, meaning the wave function extended beyond the nuclear interaction range,therefore, excited state was not expected [1].

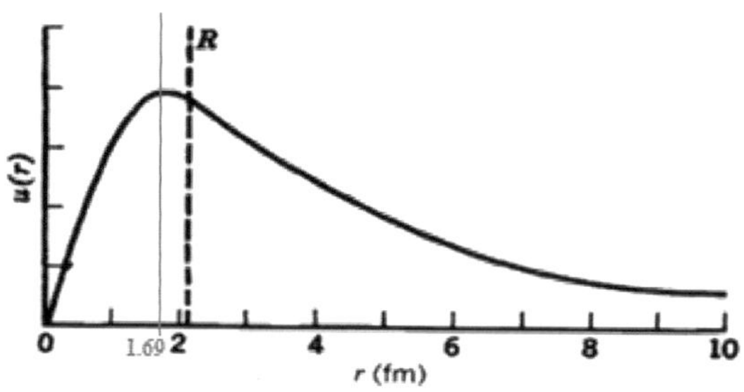

Figure 3. The deuteron wave function for $R = 2.1\ fm$.

As we can remarkably, observed Figure 2 has a relationship and connection with deuteron wave function shown in Figure 3. The maximum height value of the wave function in Figure 3 coincides with the minimum value of the potential depth graph against the deuteron radius *R*.

4. Conclusion

The trancedental equation was coded into the Python to calculate the deuteron potential depth (V_o), and reasonable solution to the problem was obtained; this also means that this study was limited to solving deuteron potential depth (V_o) at the ground state based on square potential well. Compared with the other solutions from other researchers, the obtained result from this study is among the close values to the typical potential depth value (–36 MeV). The deuteron potential value that was found in this study shows the reliability of Python programming language due to its capability and simplicity.

5. References

[1] Krane S K1988 *Introductory Nuclear physic* (John Wiley & Sons, Inc).
[2] Kelkar G N, Bedoya D 2017 *Phys. Lett B*-**772** 163.
[3] Nakayama S, Watanabe Y 2016 *J.. Nucl. Sci. Technol.* **53** 101.
[4] Chintalapati CV2013 *J. Mod. Phys.* **9** 1180 -1184.
[5] Hernandez J O, Ekström A, Nevo Dinur N, *et al.* 2018*Phys. Lett B.* **778** 383.
[6] Pohl R, Nez F, Fernandes L, *et al.* 2016 *Science* **353** 673.
[7] Zhaba I V 2018 *Mod. Phys. Lett A.* **27** 13.
[8] Sudhakar S D 2016*Am. J. Mod. Phys.***5** 46 - 55.
[9] Potel G, Perdikakis G, Atkinson M,*et al.* 2017*Eur. Phys. J. A.***53** 178.
[10] Krane S K1982 *Am. J. Phys.***50** 521-526.
[11] Schild A, Fornt. 2019 *Chem.* **79**.
[12] Sprung D W, Van Dijk W, Wang E, *et al.*1994 *Phys. Rev. C.***49** 2942-2949.
[13] Leon V D 2019 Lecture on Quantum mechanics for Engineers, www.eng.fsu.edu/~dommelen/quantum/.
[14] Burge E J, Eur. 1985 *Phys. J.* **71** 39-141.
[15] Paola C 2012 Lecture note in Introduction to Applied Nuclear Physics: Nuclear structure, MIT Open Course Ware. MIT Course Number 22.02. https://ocw.mit.edu/courses/nuclear-engineering/22-02-introduction-to-applied-nuclear-physics-spring-2012/#
[16] Edward J G 2012 Problems solutions, sol07. http://grothserver.princeton.edu/~groth/phy505f12/sol07.pdf
[17] Lembra Y, Nauk A I 1982 *Izvestiya Akademii Nauk Ehstonskoj SSR, Fizika, Matematika* **31** 345-346.

Acknowledgments
The authors would like to thank International University of Africa (IUA) for providing facilities. Special thanks to Dr. Mustapha Ahmad and Dr. Hafiz Ibrahim for their unlimited support.

A comparative photon shielding properties of protective Window materials by using EGS5 code

Nurhakimah Mamat[1], Suffian Mohamad Tajudin[1,(a)], Adila Hanim Aminordin Sabri[1], Arif Faddilah Mohd Noor[1], Nursyafika Norizan[1] and Mohd Zahri Abdul Aziz[2]

[1] Faculty of Health Sciences, Universiti Sultan Zainal Abidin (UniSZA), Terengganu, Malaysia.

[2] Oncological and Radiological Science Cluster, Advance Medical and Dental Institute, Universiti Sains Malaysia, Penang, Malaysia.

[a]suffian@unisza.edu.my

Abstract. For an X-ray room, there is a special window for radiation workers to monitor the patient during an irradiation procedure. Lead glass commonly used as protective window materials for diagnostic X-rays even it is known for its toxicity due to the lead oxide in the air which could be inhaled either during manufacturing or recycled process. Monte Carlo simulation (EGS5 code) was used to study several other types of available protective windows for its radiation shielding efficiency. In this study, the practicability of acrylic glass, plate glass and Pyrex glass systems with respect to lead glass were studied for protective window materials. The photon attenuation coefficients (μ), the percentage of transmitted and reflected photon energy spectra and half-value layer (HVL) were investigated by using simulation. Three photon source energies of 59.5 keV, 150 keV, and X-ray 150 kVp were calculated via EGS5 code to study the capability of these four glass systems as low energy photons radiation shielding materials. A thickness of 7 cm plate glass, 8 cm pyrex glass and more than 10 cm acrylic glass are adequate as a non-lead shielding material for 150 kVp X-ray to have a similar lead-glass equivalency. The use of plate glass as radiation protecting material to replace lead glass can reduce expenses as plate is way cheaper than lead but the increase thickness of plate will need special framing and concerns of its increased weight needs to be considered. However, given that the X-ray primary beam will not be directed towards the protective windows, therefore the findings are appropriate, with less thickness of plate glass could be used.

1. Introduction

Radiation protection is a specialised area and important in the radiation and radioactive sources related working environment. Its application is not just for the patient but more importantly for the staff or personnel who works daily with radiation. Functionally, radiation shielding materials are used to reduce radiation dose exposed to staff or personnel that outside of an irradiation room [1, 2, 3, 4, 5]. The designing and construction of effective radiation shielding materials requires an in-depth knowledge regarding the types of interaction between radiation and the target material [6, 7, 8, 9, 10, 11, 12].

Apart from doors and walls, protective windows are installed in a control room near the control panel for viewing. All of them must have the same lead equivalence. Referring to Standards & Industrial Research Institute of Malaysia [13], lead equivalence is defined as the protective value of a specified thickness of any materials expressed in terms of the thickness of the lead which would provide an equivalent value, where the protective material concerned is not lead on its own or a lead compound for example PbO_2. It is stated that for general radiographic X-ray systems capable of operating at potentials above 100 kVp, the lead glass observation window required minimum thickness of 1.5 mm lead equivalence while for the size, minimal dimensions of 75 cm of width by 45 cm of height.

In this study, lead glass, acrylic glass, plate glass and Pyrex glass are used for radiation shielding materials. Lead glass has many exceptional properties for instance high density, excellent infrared transmission, high non-linear optical susceptibility and high refractive index [14]. However, the lead glass is easily damaged during handling due to its soft structure, consequently increasing the amount of repairs required [15]. Concerning the standard lead glass, it is interesting to know the required thickness for other types of glass for low energy photons. As elemental composition (Table 1) for each glass is different, the type of photon interactions such as photo electric absorption, Rayleigh scattering and Compton scattering within the material could be examined by using simulation. Lead material is relatively very expensive. As the area of protective windows is large, the study of other glass materials with reasonable thickness is beneficial, without degrading the radiation shielding effect. Lead glass due to its weakness of easily damaged, would lead to more expenses for the repairs needed.

Table 1. The elemental composition of glasses [17].

Element	**Fraction by weight**			
	Lead	**Acrylic**	**Plate**	**Pyrex**
H^1		0.0805		
B^5				0.0401
C^6		0.5998		
O^8	0.1565	0.3196	0.4598	0.5396
Na^{11}			0.0964	0.0282
Al^{13}				0.0116
Si^{14}	0.0809		0.3366	0.3772
K^{19}				0.0033
Ca^{20}			0.1072	
Ti^{22}	0.0081			
As^{33}	0.0027			
Pb^{82}	0.7519			

2. Calculated transmitted photons of the elements for 150 keV photon

EGS5 code has been developed at High Energy Accelerator Research Organization, Japan. EGS5 code is a general purpose package for the Monte Carlo simulation to transport electrons and photons in an arbitrary geometry with energies above a few keV up to several hundred GeV. For verification of our Monte Carlo (EGS5 code [16]) input model with theory and literature data, we calculated the percentage of transmitted photon after the glass for single-photon energy of 150 keV photons. The density and atomic compositions of each glass were based on NIST database [17].

Figure 1 shows the materials geometry was modelled as rectangular shape with a pencil beam of mono-energy photons was incident to the center of element. The number of histories is set at 100000 of source particles to have the statistics error less than 3 %. The distance was set at 20 cm from the surface of material. All the main photon's interaction, such as photoelectric absorption, Compton scattering (including Rayleigh scattering), and pair productions were considered in our calculation. The scattered photons that penetrate the material were ignored for comparison with the attenuation coefficients generated by XCOM.

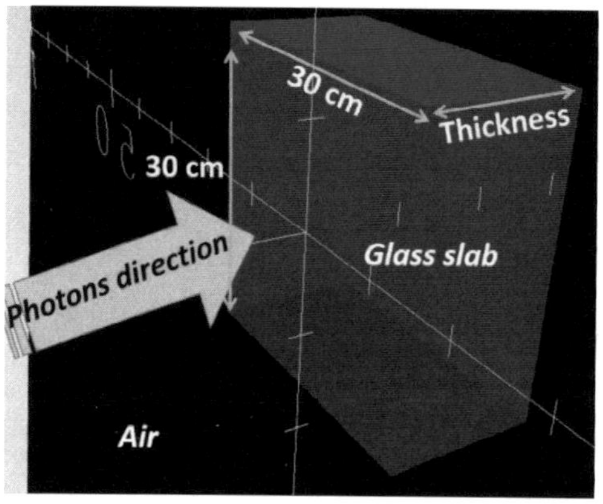

Figure 1. Simulation geometry modelled in EGS-CG view for calculation of photons attenuation.

Table 2. The percentage difference of linear attenuation coefficient for glasses at 150 keV.

Photon Energy 150 keV	**EGS5**	**XCOM**	**% of difference**
Acrylic glass	0.174	0.17	0.56
Lead glass	9.612	9.64	0.29
Plate glass	0.346	0.34	1.75
Pyrex glass	0.315	0.31	1.60

Figure 2 shows the calculation of transmitted photons as a function of material thickness for a pencil beam of 150 keV photon for each type of glass. From this graph, it can be seen that the percentage of calculated transmitted photon decreased exponentially with increasing thickness of material. Then, the data were fitted with an exponential equation to obtain the value of the linear attenuation coefficient (μ; cm^{-1}). Overall, the calculated μ value has a good agreement, within 2 % with the XCOM as in Table 2. This indicates that the μ value calculated by the simulation was reliable and verified to be used for the study of photon interactions between radiations and energy within the material.

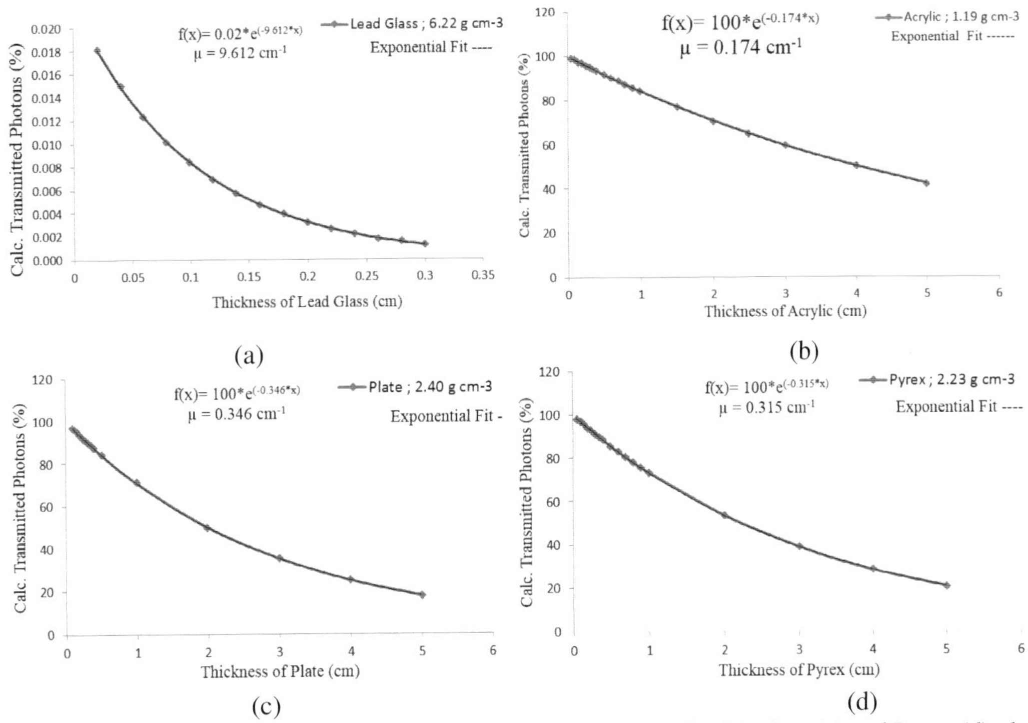

Figure 2. The calculated transmitted photons of lead glass (a), acrylic (b), plate (c) and Pyrex (d) glasses for photon of 150 keV. Calculated μ values were deduced from exponential fitting of the graph data and were compared to XCOM values.

3. The glasses attenuation coefficient for X-ray of 150 kVp

The calculation for X-ray of 150 kVp was done to see the impact of transmitted and reflected photon spectra of the same shielding glass compound. Report 78 Spectrum Processor of IPEM 1997 software [18] was used to obtain the unfiltered X-ray spectra as a source input for EGS5 code as shown in Figure 3.

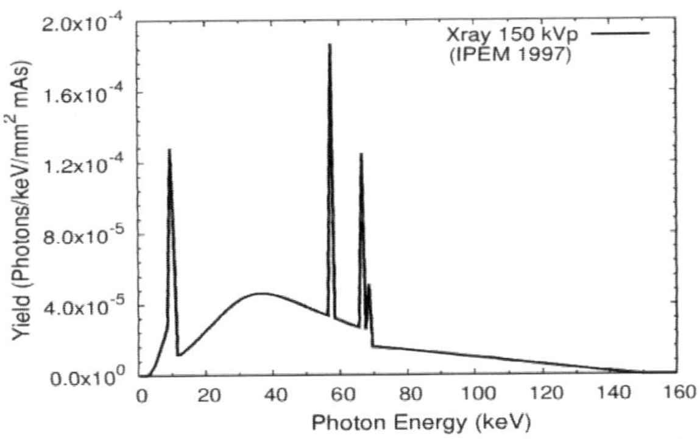

Figure 3. X-ray photon spectra of 150 kVp as a source input in the calculation. The average energy is 49.8 keV. The X-ray spectra was adopted from Report No. 78.

To calculate poly-energy X-ray of 150 kVp in EGS5 code was time-consuming. It has to be noted that practically portable Am-241 gamma source was utilised to check the irradiation room shielding before the installation of an X-ray machine. The average energy of above X-ray photon spectra was 50.7 keV. This average energy was close to Am-241's 59.5 keV. We therefore include also the calculation of 59.5 keV to estimate the difference between Am-241 and X-ray of 150 kVp.

Table 3. The linear attenuation coefficients values of lead glass, plate glass, Pyrex glass, and acrylic glass obtained by EGS5 and WinXCom for 59.5 keV and 150 kVp

Glass Materials	Linear Attenuation Coefficients, LAC (cm^{-1})		
	59.5 keV		X-ray 150 kVp
	EGS5	XCOM	EGS5
Lead glass	23.82	24.41	18.84
Plate glass	0.72	0.71	0.66
Pyrex glass	0.56	0.54	0.58
Acrylic glass	0.24	0.23	0.28

As for the glass comparison, it could be noted that lead glass had the highest LAC values for both experimental and theoretical values for energy of 59.5 keV while the acrylic glass had the lowest values of linear attenuation coefficients. The LAC results were mainly influenced by the density of material. Lead glass had the highest density while acrylic glass had the lowest density that was obtained from NIST website. The plate glass and Pyrex glass LAC values were close to each other with small amount of difference as the density for both glasses were almost similar.

In case for poly-energetic beam X-ray of 150 kVp, lead glass obviously had the highest LAC values than other three glasses mainly because of its high density, 6.22 g cm-3 which contained the element of lead with high atomic number of 82 (NIST). The high Z element contributed 75% fraction by weight in the lead glass chemical composition. According to Mohd et al. (2010, 2014), lead was a strong attenuator which had the property of attenuating X-ray radiation due to its high density which was 11.34 g/cm3. As the density or atomic number of a material increased, the attenuation produced by a given thickness would be increased (McKetty, 1998). Among the three glasses excluding lead glass, acrylic glass had the lowest LAC value because of its lowest density of glass. As shown in the Table 1, the chemical composition of acrylic glass only contained low atomic number elements which were hydrogen, carbon and oxygen. Between plate and Pyrex glass, both had almost similar density which was 2.40 and 2.23 g cm-3 (NIST,2020), respectively. However the LAC value of plate glass was significantly higher more 13.8% than the value for Pyrex glass. It has to be noted that, among the three glasses studied, plate glass contained the high Z element which was Calcium (Ca-20) within 11% in its mixture.

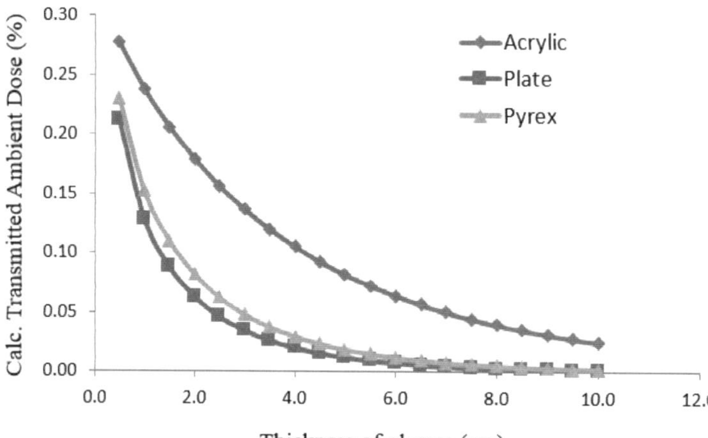

Figure 4. The calculated transmitted dose (Sv/inc./cm^2/s) for several glasses.

Among the three glasses, it could be noted that plate was better attenuator for it had less transmitted dose (%) at lower thickness compared to other two glasses, acrylic and Pyrex. It could be concluded from the graph that, 6 cm plate, 7 cm Pyrex, and more than 10 cm acrylic were needed to have the same output of lead glass at 0.02 cm thickness. A thickness of 0.12 cm lead glass, 4 cm plate, 4.5 cm Pyrex and 9.5 cm acrylic glass were adequate to reduce dose up to 90% of the photons from 150 kVp energy source. According to Direct Scientific (2020), acrylic had to be approximately five times thicker than lead glass for the same lead equivalent which significantly decreased observation capabilities. For the same size requirement and lead equivalent, acrylic had almost twice the weight of glass about 1.8 times. Nevertheless, researchers from a previous study stated that plate glass was completely clear rather than lead glass which had a yellowish colour [19].

4. Conclusion

In designing radiation shielding, to explore new element or compound materials, one must consider to account for photon energy spectra and dose whether transmitted or attenuated. The study results were useful for fabricating a lead-free radiation shielding glass in the appropriate energy range to reduce lead toxicity. It could be concluded that a thickness of 7 cm plate glass, 8 cm of Pyrex glass, and more than 10 cm of acrylic glass was adequate as a non-lead shielding material for X-ray 150 kVp in diagnostic and interventional radiology. The use of plate glass as the radiation protecting material to replace lead glass could reduce expenses as plate was cheaper than lead but the increased thickness of plate would need a special framing and concerns of its increased weight factor needed to be considered. However, given that the X-ray primary beam would not be directed towards the protective windows, therefore the findings are appropriate, with less thickness of plate glass could be used. Besides of cost, a protective window designer must take into account of other significant factors such as factor of weight, size and the requirement of special framing.

5. References

[1] Olukotun S F, Gbenu S T, Oladejo O F, Sayyed M I, Tajudin S M, Amosun A A and Fasasi M K 2020 Investigation of gamma ray shielding capability of fabricated clay-polyethylene composites using EGS5, XCOM and Phy-X/PSD *Radiat. Phys. and Chem.* **7** 109079.

[2] Abdul Aziz M Z, Yani S, Haryanto F, Ya Ali N K, Tajudin S M, Iwase H and Musarudin M 2020 Monte Carlo simulation of X-ray room shielding in diagnostic radiology using PHITS code *J. of Radiat. Res. and Appl. Sci.* https://doi.org/10.1080/16878507.2020.1828020.

[3] Kun Y, Wenyun L, Xiaoqing D, Chuanshan W, Guohua W, Mawei J, Yuanzi Z 2009 A new lead-free radiation shielding material for radiotherapy *Radiat. Prot. Dosim* **133** 256–260

[4] Aminordin Sabri A H, Abdul Aziz M Z, Olukotun S F, Tabbakh F and Tajudin S M 2020 Study on the shielding materials for low-energy gamma sources *IOP Conf. Ser.: Mater. Sci. and Eng.* **785** 012007

[5] Agar O, Sayyed M I, Akman F, Tekin H O and Kaçal M R 2018 An extensive investigation on gamma ray shielding features of Pd/Ag-based alloys *Nuclear Engineering and Technology* **51**(3) 853–859 http://doi.org/10.1016/j.net.2018.12.014.

[6] Aminordin Sabri A H and Tajudin S M 2019 Simulation of photon attenuation through Aluminium using e-radiation simulator (EGS5 Code) *J. Fundam. App.l Sci.* **1** 173–183.

[7] Tajudin S M and Aminordin Sabri A H 2019 SIMU-RAD programme: A learning tool for radiation (photons and charged particles) interaction *The J. of Polish Soc. of Med. Phys.***25** (3) 189–192.

[8] Bashter II 1997 Calculation of radiation attenuation coefficients for shielding concretes *Ann. of Nucl. Energy* **24**(17) 1389–1401 http://doi.org/10.1016/S0306-4549 (97)00003-0.

[9] Abdul Aziz M Z, Yani S, Haryanto F, Ya Ali N K, Tajudin S M, Iwase H and Musarudin M 2020 Monte Carlo simulation of X-ray room shielding in diagnostic radiology using PHITS code *J. of Radiat. Res. and Appl. Sci.* https://doi.org/10.1080/16878507.2020.1828020.

[10] Tajudin S M, Aminordin Sabri A H, Abdul Aziz M Z, Olukotun S F, Ojo B M and Fasasi M K 2019 Feasibility of clay-shielding material for low-energy photons (Gamma/X) *Nucl. Eng.and Technol.* ISSN 1738-5733.
[11] Tajudin S M and Tabbakh F 2019 Biological polymeric shielding design for an X-ray laboratory using Monte Carlo codes *Radiol.l Phys. and Technol.* **12**(3) 299–304 https://doi.org/10.1007/s12194-019-00522-w
[12] Tajudin S M, Aminordin Sabri A H, Abdul Aziz M Z and Tabbakh F 2021 Gadolinium-doped polymeric as a shielding material for X-ray *IOP Conf. Ser.: Mater. Sci. and Eng.* **1106**(1) 012010 https://doi.org/10.1088/1757-899x/1106/1/012010.
[13] Standards and Industrial Research Institute of Malaysia 1985 Malaysian Standard: MS 838. *Code of Prac. for Radiat. Protec.-Med. X-ray Diagnosis*
[14] Kaundal R S, Kaur S, Singh N and Singh K J 2010 Investigation of structural properties of lead strontium borate glasses for gamma-ray shielding applications *J.l of Phys. and Chem. of Sol.* **71**(9) 1191-1195 https://doi.org/10.1016/j.jpcs.2010.04.016.
[15] Hynes M J and Jonson B 1997 Lead, glass and the environment *Chem. Soc. Rev.* **26**(2) 133 https://doi.org/10.1039/cs9972600133.
[16] Hirayama H, Namito Y, Bielajew A F, Wilderman S J and Nelson W R 2010 Technical report SLAC-R-730 *KEK Report 2005-8: The EGS5 code system.*
[17] Gerward L, Guilbert N, Jensen K B and Levring H 2004 WinXCom-A Program for calculating X-ray attenuation coefficients *Radiat. Phys. Chem.* **71** 653-654
[18] Cranley K, Gilmore BJ, Fogarty GWA and Desponds L 1997 Catalogue of spectral data for diagnostic X-rays *report No. 78 (IPEM report 78 electronic version)*
[19] Aldrich J E and Andrew J W 1983 The design of shielding windows using lead glass, lead acrylic, and plate glass *Radiol.* **146**(1) 247-247 https://doi.org/10.1148/radiology.146.1.6849060.

Evaluation of scattering effects for radiation shielding or filter materials by using Monte Carlo simulation

Nur Miza Osman[1], Suffian Mohamad Tajudin[1,(a)], Adila Hanim Aminordin Sabri[1], Arif Faddilah Mohd Noor[1], and Mohd Zahri Abdul Aziz[2]
[1] Faculty of Health Sciences, Universiti Sultan Zainal Abidin (UniSZA), Terengganu, Malaysia.
[2] Oncological and Radiological Science Cluster, Advance Medical and Dental Institute, Universiti Sains Malaysia, Penang, Malaysia.

suffian@unisza.edu.my

Abstract. Most radiology departments utilize ordinary concrete and lead for radiation shielding as the primary radiation can be reduced through photon absorption. There are many studies done focusing on the transmitted photons that penetrate the shielding materials for radiation shielding. However, the scattering from the shielding materials would be ignored. When high-energy photons impinge on thick shields, most of the incident energy is absorbed in the shielding materials, but some of it can also be deflected sideways or in a backward direction. This is important as the backscatter radiation can contribute to unnecessary additional radiation dose to healthcare workers. Hence, this study evaluates several shielding materials namely aluminium, iron, copper, lead, ordinary concrete, and heavy concrete particularly for its attenuated and scattered photons for radiation shielding. The shielding materials were evaluated using the Monte Carlo simulation, specifically PHITS code. In the simulation, all shielding materials were modelled as a fixed 30 x 30 cm rectangular shape with a fixed thickness of 10 cm. Mono-energy and pencil beam photon energies ranging from 100 keV until 1 MeV were directed to the shielding materials. As a result, at 100 keV, lead shielding showed the least amount of transmitted dose compared to other shielding materials. However, lead shielding also showed the highest reflected dose at the same incident photon energy. As copper showed the least amount of reflected dose at this incident energy, hence applying a thin layer of this material to lead shielding can tolerate the compromise between low transmitted dose and high reflected dose. Therefore, this can improve the radiation shielding at various irradiation facilities. In conclusion, the reflected dose for all materials studied will increase or higher when the incident photon energy increase, except for lead as well as for low-Z element materials rather than high-Z element materials.

1. Introduction

Ordinary concrete and lead are commonly used as shielding materials in the radiology department to reduce the primary radiation by photon absorption or transmission. Although primary radiation can be reduced, the photons might also be scattered either by transmission or reflection from the shielding materials themselves. Many studies were done focusing on the efficacy of the shielding materials in attenuating photons for radiation shielding, specifically for the transmitted photons that passed through the shielding materials. The scattering effects from the shielding materials were often ignored, such as backscattered radiation. Backscattered radiation can be a source of unnecessary additional radiation dose to the workers that may stay inside the irradiation room. Therefore, this study not only focuses on the photon attenuation but also the scattered radiation by evaluating shielding materials that would produce the least scattered radiation by using the PHITS code to improve the radiation shielding.

Lead (Pb-82) material in the order of millimeters thickness had been widely used since 1970 to attenuate the photons either to shield the emission of photons from radioactive source or used in a large scale to shield the room [1,2]. While attenuate the photons, the lead material itself is poisonous and and expensive [3, 4,5]. Many scientists had proposed other materials to replace the lead element as a gamma or X-ray shielding such as clay [6, 7, 8, 9, 10, 11, 12, 13, 14, 15]. As an example, the authors had successfully demonstrated the polymeric compounds doped with heavy element other than lead as an alternative shielding material for 150 keV photons with an affordable thickness [16, 17].

In this study, the reflected photons and ambient dose equivalent (Sv/photon) were evaluated by Monte Carlo simulation (PHITS code [18]) for several commonly used shielding or filtering materials. This outcome enabled us to shield properly the photons that transmitted and reflected by the material in our facility.

2. Calculated Transmitted Photons of the Elements

For verification of our Monte Carlo code with theory and literature data, we calculated the percentage of transmitted photon after the elements such as Gd-64 and Pb-82 for single-photon energy of 150 keV photons. Figure 1 shows the materials geometry was modelled as rectangular shape with a pencil beam of mono-energy photons was incident to the center of element. The distance was set at 20 cm from the surface of material. All the main photon's interaction, such as photoelectric absorption, Compton scattering (including Rayleigh scattering), and pair productions were considered in our calculation. The scattered photons that penetrate the material were ignored for comparison with the attenuation coefficients generated by XCOM [19].

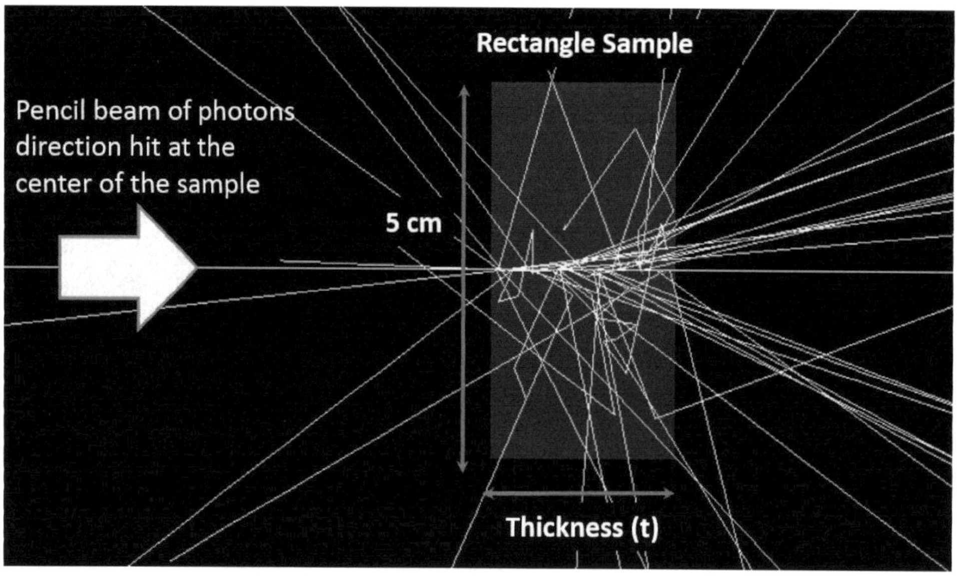

Figure 1. Simulation geometry modeled in simulation for estimating the effect of elements on the transmitted and scattered photons.

Figure 2 shows the calculation of transmitted photons as a function of material thickness for a pencil beam of 150 keV photon. From this graph, it can be seen that the percentage of calculated transmitted photon decreased exponentially with increasing thickness of material. Then, the data were fitted with an exponential equation to obtain the value of the linear attenuation coefficient (μ; cm^{-1}). Overall, the calculated μ value has a good agreement, within 1 % with the XCOM as in Table 1. This indicates that the μ value calculated by the simulation was reliable and verified to be used for the study of photon interactions between radiations and energy within the material.

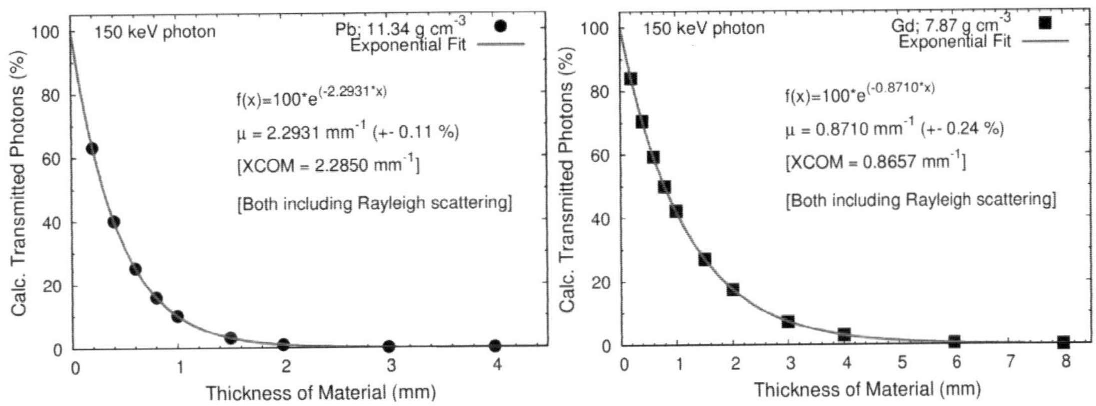

Figure 2. The calculated transmitted photons of Pb-82 and Gd-64 for photon of 150 keV. Calculated μ values are 22.93 cm^{-1} and 8.694 cm^{-1} for Pb-82 and Gd-64, respectively.

Table 1. The percentage difference of linear attenuation coefficient for Gd-64 and Pb-82 at 150 keV.

150 keV	**Linear Attenuation Coefficient (μ; cm^{-1})**		
	PHITS	**XCOM**	**% of Difference**
Gd-64	8.694	8.690	0.05
Pb-82	22.931	22.839	0.40

3. The Transmitted and Reflected Photon Dose of Several Materials

It is necessary to understand the source and mechanism through which the radiation is attenuated. This is important when determining the best shielding material for that particular radiation as several factors will influence the effectiveness of shielding such as the energy of the radiation and the atomic number of the shielding element. Figure 3 shows a graph of the transmitted dose as a function of incident photon energy from 100 keV to 1 MeV. Based on the graph, it can be seen that as the incident photon energy increased, the transmitted dose also increased for all shielding materials. This means that the higher the energy of the incident photon, the lower the amount of photon that can be trapped or absorbed by the shielding materials.

At 600 keV, the transmitted dose from Al-13 is higher within 13% compared to ordinary concrete. Heavy concrete is significantly effective to attenuate photons compared to ordinary concrete particularly at low energy region. At 400 keV, the transmitted dose from heavy concrete is lower about 77% compared to ordinary concrete while at 1 MeV, the transmitted dose from heavy concrete is 42% lower than ordinary concrete. In the case of an element, Fe-26 and Cu-29 show almost similar ability in photon attenuation as their Z number are close each other. At 500 keV, the transmitted dose for Cu-29 is 100% lower than Fe-26 while at 1 MeV, the transmitted dose for Cu-29 is 36% lower than Fe-26. Lead has the least amount of transmitted dose among other materials as it has the highest atomic number. As photoelectric effects is more dominant in materials with higher atomic number, lead can trap more photons than other materials. Hence reducing the number of transmitted photons.

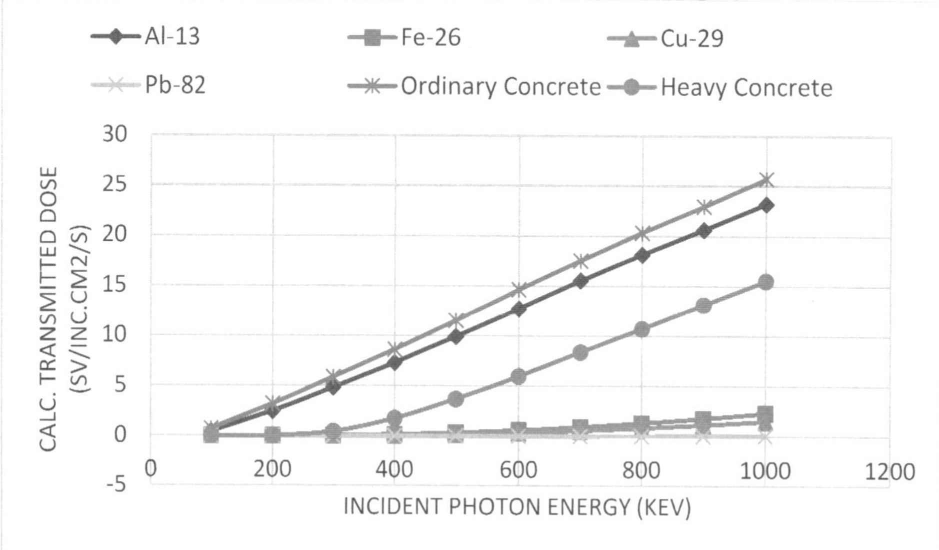

Figure 3. The calculated transmitted dose (Sv/inc./cm^2/s) as a function of incident photon energy for several elements.

Reflected photons are scored as a result of Compton scattering photons with a large angle within the materials. The probability occurs the scattered photons depends on the incident photon energy and the atomic number of the materials. As the reflected scattered photons can be a source of unnecessary additional radiation dose to the patients and workers that may stay inside the irradiation room, or close to the shielding material, therefore it is important to evaluate the reflected photons in terms of their dose and energy. Figure 4 shows a graph of reflected dose as a function of incident photon energy for shielding materials Al-13, Fe-26, Cu-29, Pb-82, ordinary concrete, and heavy concrete. Based on the graph in Figure 4, it can be seen that the reflected dose from all studied materials increased when the incident photon energy increased, except for Pb-82.

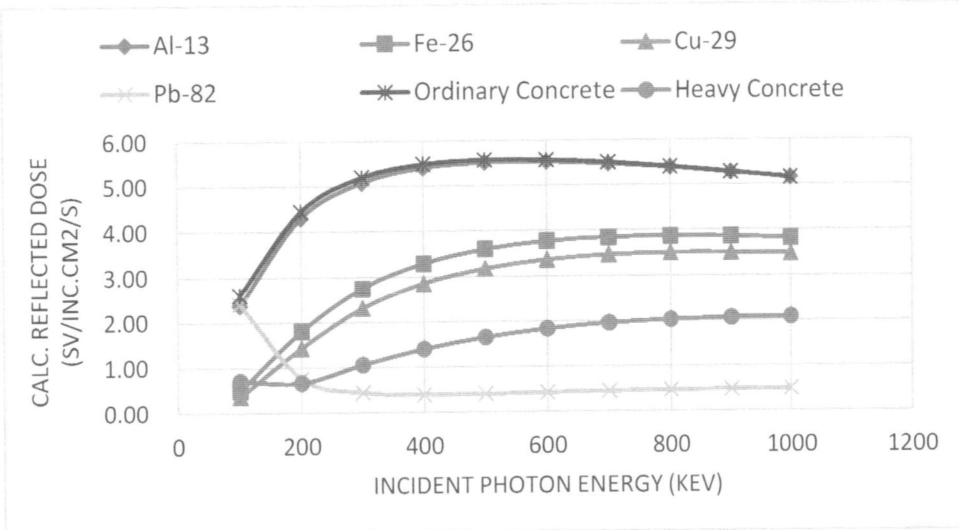

Figure 4. The calculated reflected dose (Sv/inc./cm2/s) as a function of incident photon energy for several elements.

In general, the reflected dose from all studied materials will increase as the incident photon energy increase, except for Pb-82. Although Pb-82 has the least transmitted dose, but at 100 keV, Pb-82 show the highest amount of reflected dose among the materials, which almost similar to Al-13 and ordinary concrete. This is because, low energy photon is unable to penetrate far within the lead material, which means most of photon interactions will occur only at the surface of lead. Therefore, the Compton scattered photons could easily exit the lead material as reflected photons. It was expected that the reflected dose from lead material will be higher for lower incident photon energy. Low Z element such as Al-13 and ordinary concrete, which commonly used as a shielding material show higher reflected dose among other material, in wide energy range. As the thickness of material does not affect the intensity of reflected dose, hence a thin layer of suitable material is enough to reduce the backscatter.

4. Conclusion

In conclusion, the intensity of scattered and backscattered radiation is affected by both the atomic number of shielding materials and the energy of the incident photon. An approach of photon calculations either outside or inside of an X-ray room must be considered when designing a radiation shielding to ensure radiation safety area. When photons impinge on shielding materials, most of the incident photon energy is absorbed within the material. But some photons at certain intensity and energy value could be scattered or back-scattered after undergoing Compton scattering interaction inside the material. The shielding materials must be studied both for its transmitted and reflected photons particularly if the shielding compound has a higher Z element or the source is higher energy photons.

For all shielding materials studied, lead has shown the best ability at attenuating photons at every incident photon energy compared to other materials. Furthermore, the reflected dose will increase or higher when the incident photon energy increases, except for lead (Pb-82). In addition, the reflected dose will also be higher for low-Z element materials rather than high-Z materials. Lead element has shown a high reflected dose in comparison to other material at 100 keV. Ordinary concrete that is commonly used as shielding material for an irradiation room shows the highest reflected dose in a wide photon energy range. From a radiation protection point of view, the reflected photon component is important to be evaluated. This is because the concrete walls closest to the radiation source could likely be the most significant contributors to the scattered radiation at the patient location. The unnecessary radiation dose given to the patient, especially workers that may stay near the shielding material may cause radiation-induced skin injuries, cataracts development, cardiovascular diseases, and radiation-induced necrosis.

References

[1] George L S, Charles F B and Ruben W P 1970 The use of lead as a shielding material *Nucl. Eng. and Design* **13** 3-145.

[2] Seung-Jae H, Ki-Jeong K, Tae-Ahn J and Hyun-Jib K 2016 Efficiency of lead aprons in blocking radiation – how protective are they? *Heliyon* **2** (5).

[3] Wani A, Ara A and Usmani J 2015 Lead toxicity: A review *Interdisciplinary Toxicology* **8** (2) 55-64.

[4] Billen P, Leccisi E, Dastidar S, Li S, Lobaton L, Spatari S and Baxter J B 2019 Comparative evaluation of lead emissions and toxicity potential in the life cycle of lead halide perovskite photovoltaics *Energy* **166** 089–1096 http://doi.org/10.1016/j.energy.2018.10.141

[5] Flora G, Gupta D and Tiwari A 2012 Toxicity of lead: A review with recent updates Interdisciplinary Toxicology **5** (2) p 47-58.

[6] Olukotun S F, Gbenu S T, Oladejo O F, Sayyed M I, Tajudin S M, Amosun A A and Fasasi M K 2020 Investigation of gamma ray shielding capability of fabricated clay-polyethylene composites using EGS5, XCOM and Phy-X/PSD *Radiat. Phys. and Chem.* **7** 109079.

[7] Abdul Aziz M Z, Yani S, Haryanto F, Ya Ali N K, Tajudin S M, Iwase H and Musarudin M 2020 Monte Carlo simulation of X-ray room shielding in diagnostic radiology using PHITS code *J. of Radiat. Res. and Appl. Sci.* https://doi.org/10.1080/16878507.2020.1828020.

[8] Kun Y, Wenyun L, Xiaoqing D, Chuanshan W, Guohua W, Mawei J, Yuanzi Z 2009 A new lead-free radiation shielding material for radiotherapy *Radiat. Prot. Dosim.* **133** p 256–260

[9] Aminordin Sabri A H, Abdul Aziz M Z, Olukotun S F, Tabbakh F and Tajudin S M 2020 Study on the shielding materials for low-energy gamma sources *IOP Conf. Ser.: Mater. Sci. and Eng.* **785** 012007.

[10] Agar O, Sayyed M I, Akman F, Tekin H O and Kaçal M R 2018 An extensive investigation on gamma ray shielding features of Pd/Ag-based alloys *Nucl. Eng. and Tech.* **51**(3) 853–859. http://doi.org/10.1016/j.net.2018.12.014

[11] Aminordin Sabri A H and Tajudin S M 2019 Simulation of photon attenuation through Aluminium using e-radiation simulator (EGS5 Code) *J. Fundam. Appl. Sci.* **1** p 173–183

[12] Tajudin S M and Aminordin Sabri A H 2019 SIMU-RAD programme: A learning tool for radiation (photons and charged particles) interaction *The J. of Polish Soc. of Med. Phys.* **25** (3) 189–192.

[13] Bashter II 1997 Calculation of radiation attenuation coefficients for shielding concretes *Ann. of Nucl. Energy* **24** (17) http://doi.org/10.1016/S0306-4549 (97)00003-0 1389–1401.

[14] Abdul Aziz M Z, Yani S, Haryanto F, Ya Ali N K, Tajudin S M, Iwase H and Musarudin M 2020 Monte Carlo simulation of X-ray room shielding in diagnostic radiology using PHITS code *J. of Radiat. Res. and Appl. Sci.* https://doi.org/10.1080/16878507.2020.1828020.

[15] Tajudin S M, Aminordin Sabri A H, Abdul Aziz M Z, Olukotun S F, Ojo B M and Fasasi M K 2019 Feasibility of clay-shielding material for low-energy photons (Gamma/X*) Nucl. Eng. and Technol.* ISSN 1738-5733.

[16] Tajudin S M and Tabbakh F 2019 Biological polymeric shielding design for an X-ray laboratory using Monte Carlo codes *Radiol. Phys. and Technol.* **12**(3) https://doi.org/10.1007/s12194-019-00522-w 299–304

[17] Tajudin S M, Aminordin Sabri A H, Abdul Aziz MZ and Tabbakh F 2021 Gadolinium-doped polymeric as a shielding material for X-ray *IOP Conf. Ser.: Mater. Sci. and Eng.* **1106**(1) https://doi.org/10.1088/1757-899x/1106/1/012010 012010.

[18] Sato T, Iwamoto Y, Hashimoto S, Ogawa T, Furuta T, Abe S. Ichiro Kai T, Tsai P E, Matsuda N, Iwase H, Shigyo N, Sihver L and Niita K 2018 Features of Particle and Heavy Ion Transport code System (PHITS) version 3.02 *J. Nucl. Sci. Technol.* **55** https://doi.org/10.1080/00223131.2017.1419890

[19] Gerward L, Guilbert N, Jensen KB and Levring H 2004 WinXCom-A Program for calculating X-ray attenuation coefficients *Radiat. Phys. Chem.* **71** 653-654

Acknowledgement

We would like to express our gratitude to Universiti Sultan Zainal Abidin (UniSZA) for supporting and funding this research under Dana Penyelidikan Universiti (DPU) Research Grant (UniSZA/2020/DPU2.0/07).

IOP Conf. Series: Materials Science and Engineering 1231 (2022) 012008 doi:10.1088/1757-899X/1231/1/012008

Physical properties of polyvinyl alcohol (PVAL) gel materials as phantoms for mammography

Franca Oyiwoja Okoh[1], Mohd Fahmi Mohd Yusof[2, (a)], Norlaili Ahmad Kabir[3], Siti Nor Azizah Abdullah[4]

[1,3,4]School of Physics, Universiti Sains Malaysia, 11800 Penang, Malaysia.
[2]School of Health Sciences, Universiti Sains Malaysia, 16150 Kelantan, Malaysia.

[a]mfahmi@usm.my

Abstract. This study evaluates the possibility of using polyvinyl alcohol (PVAL) composites as phantom material to simulate different densities of breast tissue for mammography based on their elastic properties. Three categories of PVAL gels were prepared by freezing and thawing 10 % water based PVAL (PVAL), ethanol and water based PVAL (PVAL/E) and a blend of PVAL with graphite powder (PVAL/G) solutions to mimic various breast densities. Young's modulus (YM) of all samples was evaluated by puncture test using the texture analyser. The results recorded ranged from 0.93 to 4.79 kPa for all samples. The results indicate that the PVAL based phantoms examined could be used to simulate fatty and dense breast tissues based on the values of their densities and YM.

1. Introduction

Phantom materials in medical physics works are used to simulate the physical, mechanical and radiological properties of the human soft tissues of specific organs. Young's modulus (YM) is a measure of the stiffness of materials and this parameter has been used as the measure of the elasticity of soft tissues [1,2]. This study focused on the development of compressible phantoms suitable for mammography. During mammography imaging, breast compressions are required to reduce breast thickness, limit the radiation dose and increasing the image quality. Although compression of the female breast is critical during mammography, none of the available standard mammographic phantoms have been similarly compressible [3]. Furthermore, commercial tissue mimicking phantoms constitute a very small range of tissue structures and they can have anatomical defects. Besides, their high cost makes it impossible for them to be used routinely as training tools. The evolution of compressible mammographic phantoms would promote the understanding of tissue deformation and verify related models.

Several studies were conducted on the characterization of biomechanical properties of soft tissues such as breast that do not have active mechanical functions but could be subject to mechanical activities like compression during mammography examination [4,5]. Different methods have been employed in the study of YM by many researchers that used the indentation technique where specimens were indented and the force-displacement slope was converted to the tissues YM [2,5,6]. The elasticity and radiological properties of solid gel samples made from 5 to 20 % PVAL concentration in equal amount of ethanol and water solution had been studied in previous works [3,7]. These were conducted by the mean of mechanical testing modalities to perform compression of samples between parallel plates while evaluating YM values at different strain levels. The results indicated that the YM of the PVAL gels ranged between 20 and 220 kPa at 15% strain level, increasing with PVAL concentrations. These values are similar to the YM of normal and cancerous breast tissue, respectively. Their result also revealed that the linear attenuation coefficients of these gels range from 0.76 to 0.86 cm^{-1} at 17.5 keV consistent with published values of breast tissue at this energy, 0.8–0.9 cm^{-1}. In our earlier study, the mass attenuation coefficients of 5 to 20 % water based PVAL solid gel samples and found them suitable for use to mimic various categories of breast tissue based on the Breast Imaging and Recording Data System (BIRADS) classification [8]. A Texture Analyser is a texture measurement system that moves in vertical directions to compress or stretch samples. It is commonly used to carry out compression, extension, puncture and rheology test in industrial food and pharmaceutical samples [9]. This study employed the use of the texture analyser to investigate the YM of PVAL composites that could mimic different categories of breast tissues based on their mass densities.

2. Methodology

2.1. Fabrication of gel samples

The PVAL granules used for this research was purchased from Sigma – Aldrich, St. Louis, MO, US. It has an average molecular weight of 85,000 – 124000 and 99+ % degree of hydrolysis. The method of preparation of the PVAL gels was in line with previous studies [3,7,8,10]. The concentration of PVAL solution used throughout this study was 10 wt %. Three categories of PVAL gels were prepared to mimic various breast densities: Water based PVAL (PVAL), ethanol and water based PVAL (PVAL/E) and a blend of PVAL with graphite powder (PVAL/G). The PVAL solution was prepared by dissolving the pre-determined amount of PVAL granules in deionised water (DI) at 90°C. The mixture was stirred continuously during heating with a magnetic stirring bar for at least 2 hours until the entire granules dissolved in water. Ethanol solution (50 %) and graphite powder (4 %) were separately added to 10 % PVAL solution to prepare PVAL/E and PVAL/G solutions respectively. The heating process was carried out in a closed system to avoid loss of liquid. The solutions were allowed to cool down at room temperature for approximately 2 hours. This process allowed the trapped air bubbles in the solution to be eliminated. The PVAL solutions were poured into cylindrical plastic moulds of known volume (height = 20 mm, diameter = 20 mm) and placed in a freezer at approximate temperature of -26°C. The gels were frozen for 12 hours and thawed at room temperature for another 12 hours to obtain solid gels. Six replicates were formed for each type of the PVAL composites evaluated: PVAL, PVAL/E and PVAL/G. Three of each type underwent 1 FTC while the remaining 3 underwent 2 FTCs.

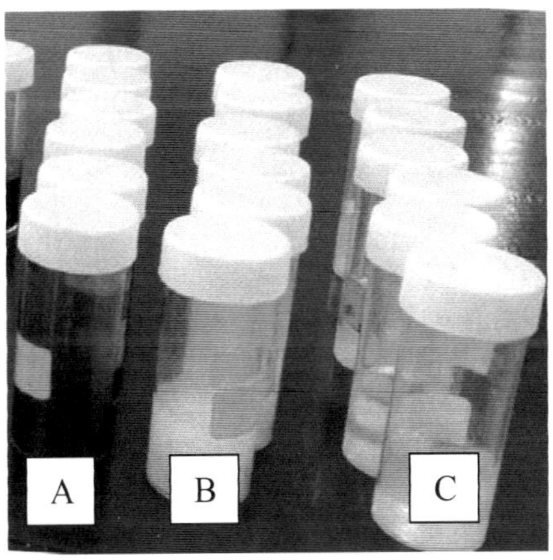

Figure 1. The fabricated PVAL composites test samples of (a) PVAL-G (b) PVAL and (c) PVAL-E samples.

2.2. Physical and Mechanical testing of gel samples

The mass density, ρ of each sample was calculated as the ratio of mass to volume (g/cm^3). The mass and volume of each sample were measured using the electronic balance and displacement method respectively. The values of 3 replicates for each sample were averaged and taken to attain a high measurement accuracy to the nearest 0.01 g/cm^3.

To characterize tissue or material stiffness, the mechanical property of PVAL composites were studied through the puncture test. Specifically, the Young's modulus (YM) was evaluated. The puncture test was carried out using the texture analyser (Model TA. XT. Plus Texture Analyzer, UK) with a 5 kg load cell. The fabricated gel samples were maintained at room temperature 24 hours prior to the test. In this test, the resistance of material to an applied force was measured. A sample was placed in the mechanical testing machine and force was applied, the aluminium probe (12.6 mm diameter) was compressed into the sample to a penetration depth of 8 mm at crosshead speed of 0.5 mm/s. This experiment was carried out on PVAL, PVAL/E and PVAL/G samples that underwent 1 and 2 FTCs to measure the YM and the effect of force on sample type and stiffness.
Young's modulus is given by:

$$E = \delta\varepsilon \tag{1}$$

with $\sigma = F/A$ (Force/cross sectional Area) and $\varepsilon = \Delta l/l$ (Changes in the length/original length). The cross-sectional area of cylindrical objects is $A = \pi r^2$ where r is the radius of the cylinder. The force-distance data was collected from TA system and YM calculated using equation 1.

3. Results and Discussion

The evaluated mean densities of the PVAL gel samples is summarised in Table 1. The result revealed that all the PVAL samples have densities approximately equal to the density of water (1.0 g/cm^3) with PVAL/E having the least density of 0.952 ± 0.011 g/cm^{3}and PVAL/G having the highest density of 1.081 ± 0.02g/cm^3. The addition of graphite powder which has a density of 2.0 g/cm^3 increased the density of PVAL/G to higher than 1.0 g/cm^3 while ethanol solution (ρ = 0.78 g/cm^3) reduced the density below water. The nearness of the mass densities of evaluated PVAL composites to the density of water and breast tissue (1.02 g/cm^3) suggests that the samples possess water/tissue equivalent properties which makes it suitable for use and breast phantom material.

Table 1. The calculated average mass density of fabricated PVAL gel samples by using the gravimetric method

Sample	Mean Density ± SD
PVAL	1.056 ± 0.002
PVAL/E	0.952 ± 0.011
PVAL/G	1.081 ± 0.02

Figure 2 shows the calculated YM (a) and Force applied (b) for PVAL, PVAL/E and PVAL/G samples that underwent 1 and 2 FTCs. The value of YM obtained for all samples ranged from 0.93 to 4.79 kPa with PVAL (1 FTC) and PVAL/G (2 FTC) recording the least and highest YM respectively. The results confirm the linear relationship that exist between force and YM. Stiffer samples required higher force to push the aluminium probe through the samples hence recoded higher YM, the reverse was the case for less stiff samples. This is because graphite powder fills the void spaces in the PVAL/G matrix thereby making the gel more compact and requiring more force to plunge through. For all categories of samples examined, the force and YM increased with an increase in the number of FTCs as expected.

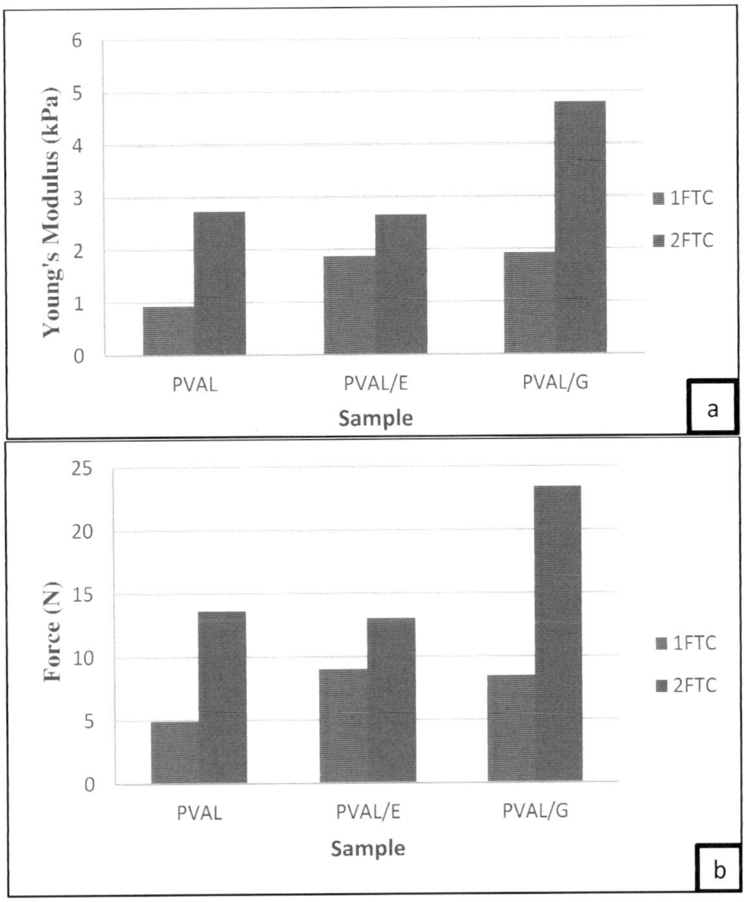

Figure 2. Calculated (a) Young's Modulus and (b) Applied Force of the PVAL, PVAL/E and PVAL/G that underwent 1 and 2 FTCs.

There is a notable disagreement between our measured YM and those of [3]. Nevertheless, our results are close to the results obtained by [2] who measured YM of agar and gelatine and obtained 5.6 – 12.2 kPa. Our results also agree with the values of YM achieved by [5] when they measured the YM of 168 breast tissue specimens in vitro and obtained 3.25 and 3.24 kPa for adipose and fibro glandular tissues respectively. Furthermore, in vivo measurements by [11,12] covers a range of 1.0 to 20.0 kPa and 2.0 to 37.0 kPa for adipose and fibro glandular tissues respectively. This can be associated with the fact that different authors used various strain levels, inconsistent sample geometry and different measurement methods. It should be remembered that differences in the YM of different tissues can span an extremely large dynamic range of elasticity [4] so even with large errors, differential diagnosis based on elasticity imaging may still be useful. In line with previous studies, the result of this study indicates that all gel samples studied could simulate different categories of breast tissue based on their density.

4. Conclusion

This study focused on the measurement of the physical and mechanical properties of PVAL- based solid gel as breast tissue equivalent material. The densities of the fabricated PVAL gel phantoms showed close resemblance to those of water and breast tissue based on AAPM, ICRU-44 1989. The YM obtained in this experiment for all samples ranged from 0.93 to 4.79 kPa. The results indicate that the PVAL based phantoms examined could be used to simulate fatty and dense breast tissue based on the values of their densities and YM.

5. Acknowledgement

The authors would like to acknowledge the financial support from Universiti Sains Malaysia under FRGS grant number USM FRGS/1/2018/STG02/USM/03/3/6711631 and USM Research Universiti (RU-Individual) Grant no. 1001/PPSK/8012381. The authors also acknowledge the School of Industrial Technology, Universiti Sains Malaysia for the use of texture analyser in this research.

6. References

[1] Ramiao N G, Martins P S, Rynkevic R, Fernandes A A, Barroso M, Santos D C. 2016 Biomechanical properties of breast tissue, a state-of-the-art review *Biomech. Model Mechanobiol.* **15**(5) p1307–1323.

[2] Samani B J, Luginbuhl C, Plewes D B A. 2003 Measuring the elastic modulus of ex vivo small tissue samples. *Phys. Med. Biol.* **48** p2183–2198.

[3] Price B D, Gibson A P, Tan L T, Royle G J. 2010 An elastically compressible phantom material with mechanical and x-ray attenuation properties equivalent to breast tissue. *Phys. Med. Biol.* **55**(4) p1177–88 Available from: https://www.ncbi.nlm.nih.gov/pubmed/20107248.

[4] Fung Y C. 1993 Mechanical properties of living tissues *Springer Book* **547**

[5] Samani A, Zubovits J, Plewes D. 2007 Elastic moduli of normal and pathological human breast tissues: an inversion-technique-based investigation of 169 samples *Phys Med Biol* **52** (6) 1565.

[6] Krouskop T A, Wheeler T M, Kallel F, Garra B S, Hall T 1998 Elastic Moduli of Breast and Prostate Tissues Under Compression *Ultrason Imaging* **20** p260–74.

[7] Price B D, Gibson A P, Royle G J 2010 Dual-modality imaging of a compressible breast phantom with realistic optical and x-ray properties *Medical Imaging 2010 Physics of Medical Imaging. International Society for Optics and Photonics* **7622** (762207).

[8] Okoh F O, Ahmad K N, Mohd Yusof M F, Abdullah A S N. Measurement of mass attenuation coefficient of polyvinyl alcohol (PVAL) as breast tissue equivalent material in the photon energy range of 16.61-25.26 keV 2020 *J. Phys. Conf. Ser.* **1535** (012051).

[9] Brenner T, Hayakawa F, Ishihara S, Tanaka Y, Nakauma M, Kohyama K 2014 Linear and Nonlinear Rheology of Mixed Polysaccharide Gels. Pt. II. Extrusion, Compression, Puncture and Extension Tests and Correlation with Sensory Evaluation. *J. Texture Stud.* **45**(1) 30–46.

[10] Ossati M S, Katy Szczepura P H 2015 Design, development and use of a deformable breast phantom to asses the relationship between thickness and lesion visibility in full field digital mammography PhD thesis University of Salford

[11] Sinkus R, Lorenzen J, Schrader D, Lorenzen M, Dargatz M, Holz D 2000 High-resolution tensor MR elastography for breast tumour detection. *Phys. Med. Biol.* **45**(6)1649

[12] Kruse S A, Smith J A, Lawrence A J, Dresner M A, Manduca A, Greenleaf J F *et al.* 2000 Tissue characterization using magnetic resonance elastography: preliminary results *Phys. Med. Biol.* **45**(6)1579.

Superconducting properties of $(Bi,Pb)_2Sr_2Ca_2Cu_3O_x$ (BSCCO-2223) superconductor ceramics prepared by conventional solid-state reaction and co-precipitation methods

Zaahidah 'Atiqah Mohiju[1], Madihah Mujaini[1], Nasri A. Hamid[1,(a)]

[1]Department of Mechanical Engineering, College of Engineering, Universiti Tenaga Nasional, Jalan IKRAM-UNITEN, 43000 Kajang, Selangor, Malaysia

[a]Nasri@uniten.edu.my

Abstract. The $(Bi,Pb)_2Sr_2Ca_2Cu_3O_x$ (BSCCO-2223) samples were prepared by conventional solid-state reaction and co-precipitation methods. The crystal size, cell parameter, and crystal structure were identified from X-ray diffraction (XRD). The peaks drawn by co-precipitation have higher intensities and the width of peaks were narrower compared to the sample prepared by solid-state method. Slight difference in the value of lattice parameters between the samples were also noticeable. Concentration of defects increased the distortion of crystal lattice that cause the changes in these parameters. The electrical properties of the superconductors were investigated by using cryogenic four-point probe and AC susceptibility. The sample prepared by co-precipitation method was able to reach higher critical temperature, T_C at 106 K compared to the other sample, T_C at 97 K. Meanwhile Scanning Electron Microscope (SEM) micrographs for solid-state method sample showed more random grain alignment and porous structures. The microstructures and grain growth results of Bi-2223 were reflected to the heat and temperature applied during sintering process. In short, the research work reveals that co-precipitation method produces better quality of high temperature superconductor compared with solid-state method due to higher homogeneity resulted by chemical reactions.

1. Introduction

The discovery of superconductivity of BSCCO family has attracted interest of researchers from all over the world. These materials have special ability which is to carry current with zero resistance and reduce power loss during the process of transmitting electricity [1]. In electrical transmission system, superconducting wire is used to produce a transmission cable. It is proven that the distinctive properties of superconductor enable the development of high efficiency energy devices. Several others electrical power devices such as motor, generator, fault current limiter and magnetic storage system have been developed [2-4].

In order to produce high quality material in large-scale manufacturing, it will depend mainly on the process conditions for the production of the BSCCO powders. The most practical technique for the production of BSCCO powders is the solid-state reaction method. The main issue of using this method is the repetition of grinding process for several times before the sintering process. It is necessary to achieve end products with fine particle size, homogeneity, high purity and with the desired superconducting properties [5]. Then, heat treatment plays an important role in the fabrication of pure

Bi-2223 bulk samples since it will diminish several unwanted phases such as Bi-2212, Bi-2201 and Ca_2PbO_4 [6–8].

The co-precipitation method is another approach of producing BSCCO oxide powder which requires precise control of the reaction environment (pH). The powders obtained by the co-precipitation techniques have a smaller grain size and has higher homogeneity than the powders of the solid-state reaction method. Existence of impurities will directly reduce the value of T_C and J_C of the samples since these impurities are located within the grains and limit the grain connectivity [9].

The bulk samples of Bi-2223 were characterized by X-ray diffraction (XRD) and scanning electron microscopy (SEM) for their phase purity, homogeneity, grain size and orientation, while the superconducting properties were tested by ac susceptibility and four-point dc-electrical resistivity methods. Throughout this investigation, a comparison of the co-precipitation and the solid-state reaction methods regarding their process conditions and the quality of the powders produced was achieved.

2. Experimental

First specimen with nominal composition $Bi_{1.6}Pb_{0.4}Sr_2Ca_2Cu_3O_{10}$ was prepared via conventional solid-state method. Samples were made of oxide and carbonate powders, homogenously mixed according to its stoichiometric molar ratio of Bi_2O_3, PbO, Sr_2CO_3, $CaCO_3$ and CuO. The mixture was ground and heated twice in a tube furnace with free air flow at temperature of 810°C for 30 hours to remove impurities and form the Bi-2223 oxide. According to V.Garnier, the calcined powder phase assemblage is sensitive to the calcination conditions, and control of the calcination parameters is necessary to form the adequate reactive secondary phases (Bi-2212, Ca_2PbO_4, Ca_2CuO_3, CuO). These phases are necessary for the proper synthesis of Bi-2223 during the sintering step [10]. Then, the oxide powder was pressed at room temperature with pressure of 7 tonnes. Bulk sample of 13 mm in diameter and 3 mm in thickness was formed and synthesized under the same condition at 850°C for 60 hours, then furnace-cooled to room temperature.

The second specimen was produced by co-precipitation method described by Hamadneh et al. [11]. All the chemical powders were acquired from Sigma-Aldrich, USA. Then, the precursor's oxide powder was pelletized and sintered to achieve the required Bi-2223 phase. The dehydrated powder was thoroughly mixed and grounded in an agate mortar. After the first grounding process, the compound went through pre-calcine process at temperature of 730°C for 12 hours. Then, after the second grounding process, the powder was calcined at temperature of 850°C for 24 hours. Next, the powder was then pressed into pellet under hydrostatic pressure around 7 tons or 70,000 psi. The pellets were sintered at maximum heat of 850°C for 48 hours and then furnace-cooled to room temperature.

Phase identification of both specimens were conducted using a Bruker D8 Advanced X-Ray Diffractometer (XRD). Microstructure investigation was carried out using a Hitachi S3400N Scanning Electron Microscope (SEM). Superconducting properties for each sample, such as transition temperature, T_C and the critical current density, J_C were determined by using AC susceptibility technique.

3. Results and Analysis

Figure 1 and Figure 2 show XRD spectrums for Bi-2223 bulk sample prepared by solid-state and co-precipitation methods. Multiphase structure was observed in both samples which indicates the difficulty in obtaining a single-phase material. This can be attributed to the phase transition between Bi-2201, Bi-2212 phase, and also the orientation of the grains. The number of corresponding peaks and their intensities indicates that the high T_C phase is the dominant phase in both samples.

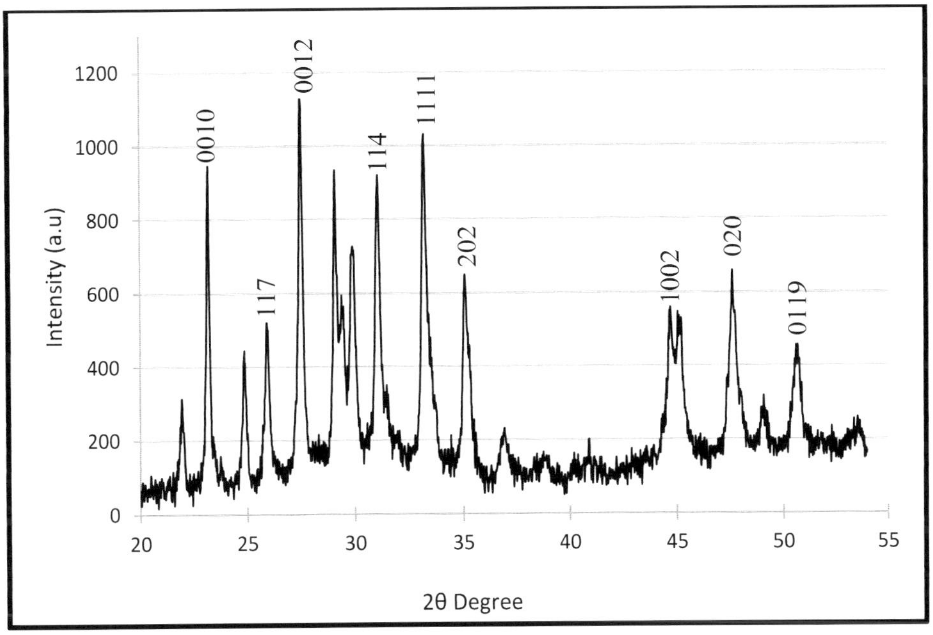

Figure 1. XRD for Bi-2223 bulk sample prepared by solid-state method.

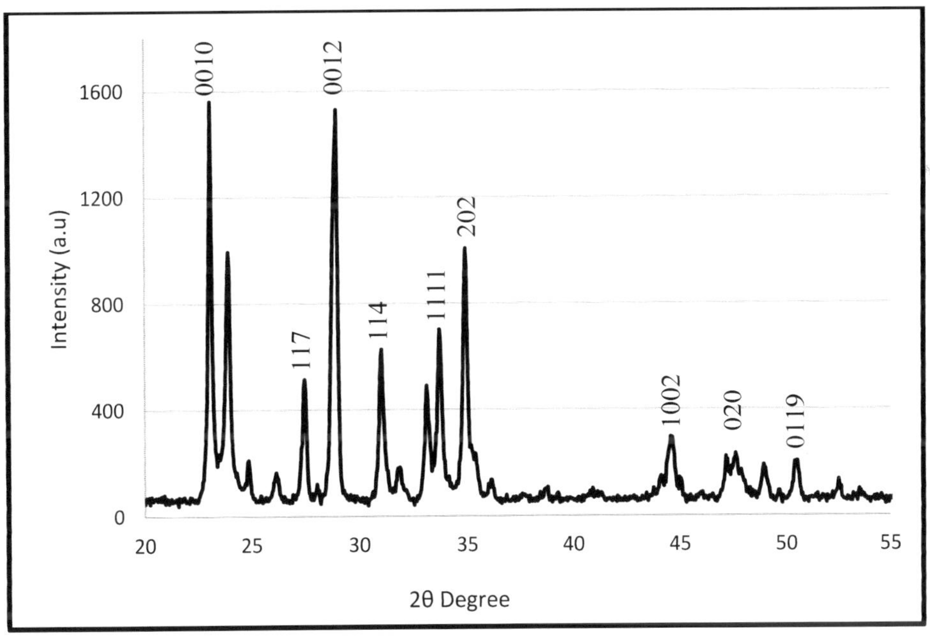

Figure 2. XRD for Bi-2223 bulk sample prepared by co-precipitation method.

Table 1 shows value of lattice parameters for both samples. The values are in agreement with the previous investigators where the parameters value obtained were still within the range [12-13]. In view of results from XRD and lattice parameters, those intense peak patterns of sample belong to the high-T_C phase, which also indicates an increase in the volume fraction of the Bi-2223 phase. Volume fraction of Bi-2223 for sample prepared by solid-state method can be increased by extending the period of sintering time.

Table 1. Lattice parameter for samples prepared by solid-state and co-precipitation methods.

Lattice Parameter	Samples	
	Solid-state Method	Co-precipitation Method
a	5.266 Å	5.408 Å
b	5.471 Å	5.408 Å
c	38.301 Å	37.015 Å

Sintering time and temperature influence Bi-2223 volume fraction. These parameters also play significant role in improving the grain coupling, the grain morphology, and the phase formation and transition, leading to the improvement of the electrical properties of Bi-2223 [14-15]. Figure 3 and Figure 4 show T_C measurement using four-point probe for the samples. Critical temperature for sample prepared by co-precipitation method was recorded at 102 K, compared to sample prepared by solid-state method which was at 95 K. Sample with higher T_C value usually has stronger link between the grains, which can be spotted later in SEM analysis by observing the grain's alignment along the c-axis. It is known that the composition variation at the grain boundaries, disoriented grain is one of the mechanisms that control the weak link behavior of polycrystalline high-T_C superconductors [16]. S.M. Cassidy also reported that the temperature and field dependence of the grain boundary weak link dissipation in BSCCO tapes drops rapidly as the temperature decreases [17].

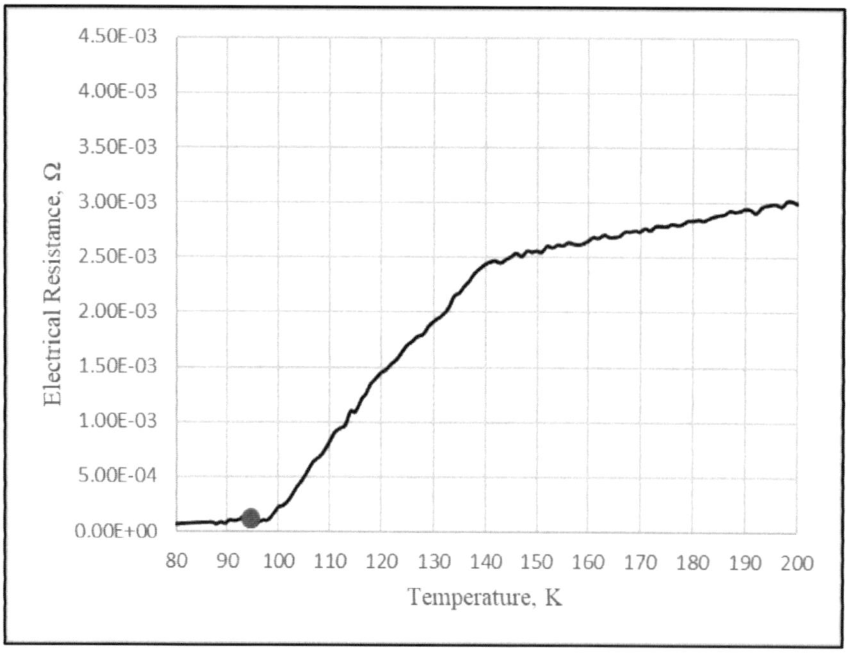

Figure 3. T_C measurement using four-point probe for Bi-2223 bulk sample prepared by solid-state method.

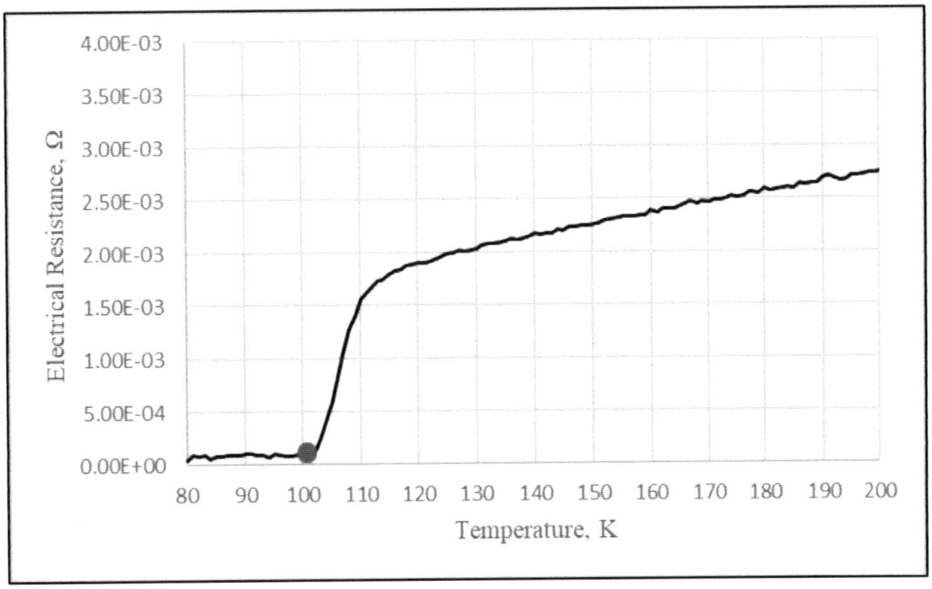

Figure 4. T_C measurement using four-point probe for Bi-2223 bulk sample prepared by co-precipitation method.

AC susceptibility has been used to measure the critical temperature and critical current of the samples. Figure 5 and Figure 6 show the real parts of susceptibility (χ') of both samples, solid-state method and co-precipitation method, respectively. In each graph, the position of inflection point on the curve is the critical temperature. This is due to the considerable difference between lower critical magnetic field of the intergranular matrix and the grains, because the flux will penetrate the intergranular matrix first before it reaches the grains [18].

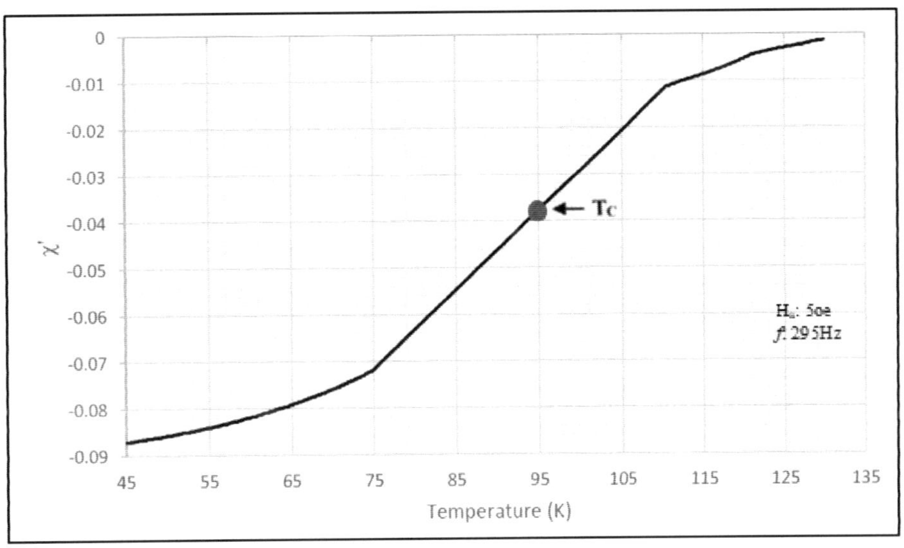

Figure 5. Real (χ') parts of susceptibility for Bi-2223 bulk sample prepared by solid-state method, applied magnetic field, H_a=5oe and frequency of 295Hz.

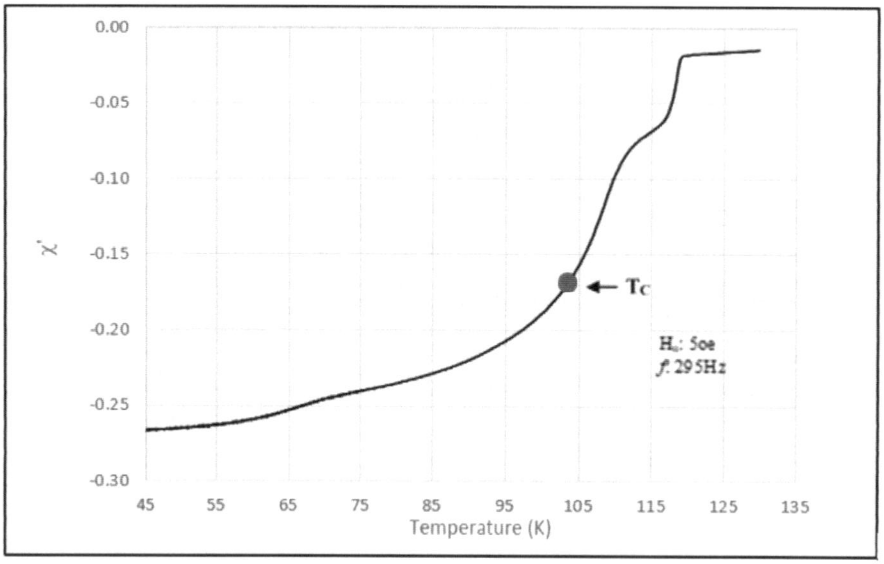

Figure 6. Real (χ′) parts of susceptibility for Bi-2223 bulk sample prepared by co-precipitation method, applied magnetic field, H_a=5oe and frequency of 295Hz.

At temperature of 77 K, critical current density, J_C for solid-state sample was recorded as 1.6176 A/cm^2, meanwhile for another sample it is 0.98 A/cm^2. Typical behaviour in high T_C superconductor materials is the increment in T_C value that promotes higher magnetic field of the material but it decreases the intra-granular J_C values.

Figures 7(a) and 7(b) show the morphologies of the prepared samples using both solid-state and co-precipitation method. There were random plate-like grains arrangements for all the samples with almost identical sizes. This structure has become a signature of the Bi-2223 phase formation from the Bi-2212 matrix mostly due to the prolonged sintering process [19]. Distribution of grains in sample prepared by co-precipitation method was more homogenous and have better grain alignment of its crystallites with c-axis. In addition, the plate-like grains were aligned to make denser and conductive sample.

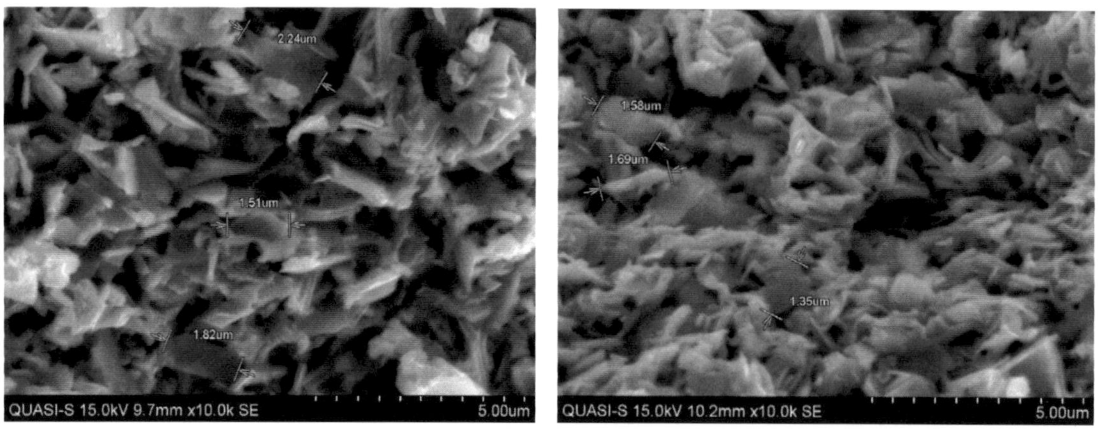

Figure 7. Scanning Electron Microscopy micrographs for Bi-2223 bulk samples prepared by (a) solid-state method and (b) co-precipitation method

4. Conclusion

The co-precipitation method has proven to provide more advantages in comparison to the solid-state reaction method with regard to the process conditions and the quality of the BSCCO powder produced. Sample prepared by the co-precipitation method is homogeneous and has better superconducting quality in comparison to the time-consuming solid-state method, which needs repetition of grinding process in order to achieve the required homogeneity.

5. References

[1] Schiferi R, Flory A, Livoti W C and Umans S 2006 High temperature superconducting synchronous motors, economic issues for industrial applications *Petroleum and Chemical Industry Conference, Industry Applications Society 53rd. Annual Meeting* (Philadelphia, Pennsylvania) 1-9.
[2] Chu C W 2000 *Physica C* **341-348** 25
[3] Noe M and Oswald B R 1999 *IEEE Trans. Appl. Supercond.* **9** 1347
[4] Friedman A, Shaked N, Perel E, Sinvani M, Wolfus Y and Yeshurun Y 1999 *Cryogenics* **39** 53.
[5] Foltyn M, Civale S R, MacManus-Driscoll L, Jia J L, Maiorov Q X, Wang B and Maley H 2007 *Nat. Mater.* **6** 631.
[6] Chen Y L and Stevens R 1992 *J. Am. Ceram. Soc.* **75** 1142
[7] Chen F H, Koo H S and Tseng T Y 1991 *Appl. Phys. Lett.* **58** 637
[8] Yoo J, Kim S, Ko J -W and Kim Y K 2004 *Supercond. Sci. Technol.* **17** S538
[9] Ghattas A, Annabi M, Zouaoui M, Ben Azzouz F and Ben Salem M 2008 *Physica C* **468** 31
[10] Garnier V, Ambrosini A and Desgardin G 2002 *J. Mater. Sci.: Mater. El.* **13** 353
[11] Hamadneh I, Halim S A and Lee C K 2006 *J. Mater. Sci.* **41** 5526
[12] Saleh S A 2006 *Physica C* **444** 40
[13] Kocabas K 1998 *Turkish J. Phys.* **22** 437
[14] Darsono N, Yoon D -H and Raju K 2016 *J. Supercond. Nov. Magn.* **29** 1491
[15] Özçelik B, Nane O, Sotelo A, Amaveda H and Madre M A 2017 *J. Mater. Sci.: Mater. El.* **28** 6278
[16] Shin D H, Silcox J, Russek S E, Lathrop D K, Moeckly B and Buhrman R A 1990 *Sci. and Technol. Thin Film Supercond.* **2** 319
[17] Cassidy S M, Cohen L F, Cuthbert M N, Dou S X and Caplin A D 1992 *Cryogenics* **32**(11) P1034
[18] Nikolo M 1995 Am. J. Phys. **63** 57
[19] Muné P, Govea-Alcaide E and Jardim R F 2003 *Physica C* **384** 491

Acknowledgements

This project is funded by the Universiti Tenaga Nasional under BOLD Research Grant reference number J510050002/2021084, and conducted under collaboration with the Malaysian Nuclear Agency. The authors wish to thank Prof. Abd-Shukor from the Universiti Kebangsaan Malaysia for providing his research facilities for this work.

Polyethylene composite with boron and tungsten additives for mixed radiation shielding

Muhammad Arif Sazali[1, a)], Nahrul Khair Alang Md Rashid[1], Khaidzir Hamzah[1], Faridah Mohamad Idris[2], Muhammad Syahir Sarkawi[1], Nur Syazwani Mohd Ali[1]

[1]School of Chemical and Energy Engineering, Faculty of Engineering, Universiti Teknologi Malaysia, 81310 Skudai, Johor, Malaysia
[2]Agensi Nuklear Malaysia, 43000 Kajang, Selangor, Malaysia

[a)]muhammadarifsazali@utm.my

Abstract. Mixed neutron and gamma radiations are common in many nuclear applications. Several materials can be combined to obtain a composite material that is better for mixed radiation than the individual component materials. The aim of this paper is to investigate the shielding effectiveness of a polyethylene (PE)-based composite with boron and tungsten additives. Several compositions are tested to shield against a ^{252}Cf fission neutron source using an attenuation experiment. The composite is also manufactured using the melt-mixing method of component raw materials. Comparisons are made between the different compositions and the experimental results. Results suggest that the PE composite with 16%wt boron and 16%wt tungsten show the best mixed radiation attenuation as compared to pure PE, PE composite with 25%wt boron, and PE composite with 25%wt tungsten.

1. Introduction

Concerns of mixed neutron and gamma radiations are common in most nuclear applications. This is due to the interaction of neutrons with matter that can result in the production of secondary gamma-rays. Therefore, the shielding for gamma-rays will also need to be considered along with the neutron shielding. Many polymer-based composites have been developed to tackle this problem [1]. This is because polymer has a high content of hydrogen which is excellent in neutron moderation. Other additives are added to improve the shielding performance of the base polymer such as boron to absorb thermal neutrons, and heavy elements to absorb secondary gamma-rays.

Boron-filled composites are popular among researchers of neutron shielding because of their high macroscopic absorption cross-section for thermal neutrons. Uddin et al. (2020) [2] investigated polyethylene(PE)/boron carbide composite for up to 40% loading. They noted that although higher compositions of boron are effective, this reduces the amount of hydrogen in the polymer base which can affect the moderation of fast neutrons. Thus, they recommended 10% boron content as the best composite formulation that balances the content of boron and hydrogen. Nurazila and Megat Harun (2020) [3] showed that adding boron carbide to a high-density PE (HDPE)/natural rubber blend improved the neutron shielding performance of the base material. However, higher boron content was found to diminish the mechanical properties of the base material. Jumpee et al. (2020) [4] also used natural rubber mixed with boron powder for neutron shielding and it exhibited good neutron attenuation. Additionally, another sample with iron oxide powder was also investigated and it produced a similar shielding performance. Although, the gamma-ray transmission was not measured which could have revealed the potential of adding the iron oxide to improve the gamma-ray attenuation of the base material.

The gamma shielding performance of a polymer-based composite can be improved by mixing heavy elements which have a higher density of electrons. Because the interaction of photons with materials usually involves electrons, heavy elements are effective in absorbing or scattering gamma-rays. Tungsten is a well-known additive element in many composite shielding studies. Alavian and Tavakoli-Anbaran (2019) [5] analyzed the effect of various sizes and proportions of tungsten particles in a low-density PE (LDPE) matrix. It was found that as the particle sizes decreased, the gamma attenuation increased. This improvement was also observed with higher proportions of the tungsten. However, the effect of particle sizes is less than that of the proportions. A similar conclusion was drawn by Asgari et al. (2020) [6] who studied the effect of tungsten and bismuth in rubber composite. Although, significant effects were only observed at higher weight percentages of the heavy elements (45-60 %). Can et al. (2021) [7] proved that regardless of the type of tungsten compounds (pure, oxide, or carbide), higher contents of tungsten are effective across a wide range of gamma-ray energies. The base polyester was made into several samples of composite with different types of tungsten compounds.

Based on the literature review, the present study aims to develop a new composite that is effective against both neutrons and gamma-rays. The base material is made from polyethylene. Boron is added to improve its neutron absorption, and tungsten is introduced to absorb gamma-rays.

2. Methods and materials

The study involves attenuation experiments with a mixed neutron and gamma-ray source. All equipment was provided by Nuclear Malaysia and Nuclear Laboratory in Universiti Teknologi Malaysia (UTM).

2.1. *Fabrication of composite shield*

The raw materials were purchased in the form of polyethylene (PE) pellets and powders of pure boron and tungsten. The new composite material was fabricated using the melt mixing method based on the configuration shown in Table 1. Firstly, the PE pellets were melted in a Haake internal mixer at 150 °C. After two minutes, the boron and tungsten powders were added. Eight minutes later, the composite mix was extracted and transferred to the hot press machine. The composite was then pressed into a 15 cm ×15 cm × 0.2 cm mold at 180 °C. After it has cooled, it was cut into 3 cm × 3 cm slabs. For dose rate measurement, five slabs were stacked together to form a 1-cm thick shield slab as shown in Figure 1.

Table 1. Composite shield configuration for production

Sample	*Composition (%wt)*		
	PE	*B*	*W*
PE100	100	–	–
B25	75	25	–
W25	75	–	25
BW32	68	16	16

2.2. *Attenuation experiment*

The shielding experiment was carried out using a Californium-252 fission neutron source in UTM. At the time of the experiment, the source strength was calculated to be 171 kBq or 4.6 μCi. Since the source is relatively weak, the shield was positioned immediately beside the source and the detector was placed 5 cm away from the source as shown in Figure 2. The background radiation was measured before the experiment to obtain the true value of the dose rate measurements.

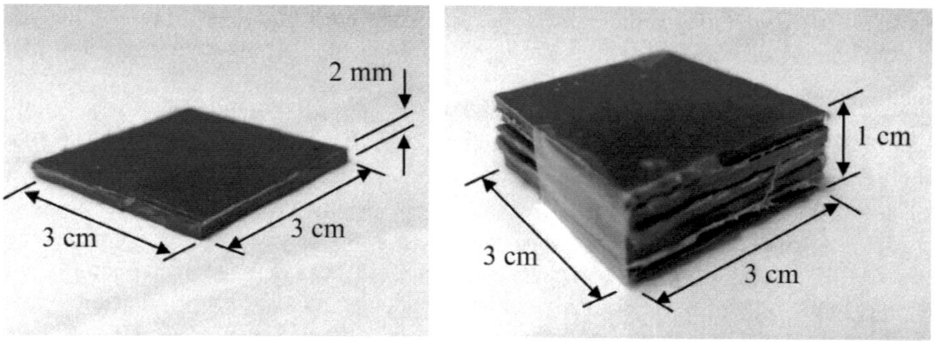

(a) (b)

Figure 1. (a) A single composite slab and (b) a 1-cm thick composite slab.

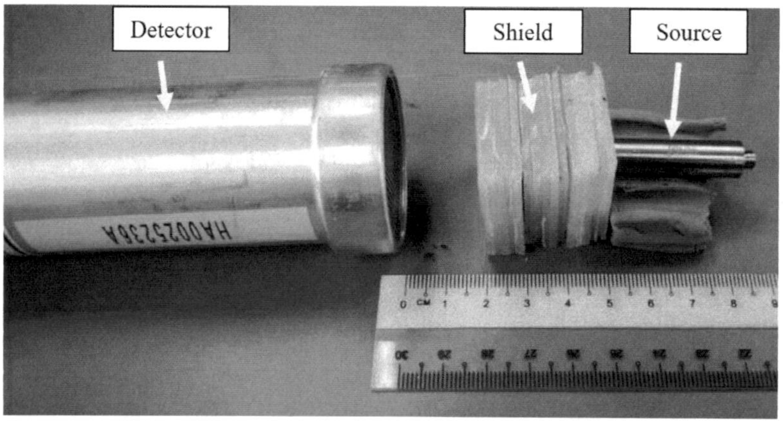

Figure 2. Experimental setup with 3 cm of PE100.

3. Results and discussion

3.1. Physical density of samples

Theoretical density has upper bound and lower bound values which were calculated using Equations (1) and (2) respectively [8]. Table 2 shows the comparison between the theoretical density and the actual density of all samples. The actual density was obtained by measuring the weight of each sample using a digital balance and dividing the weight by the volume of the sample. It can be observed that the B25 and W25 composites have an actual density that is within the range of the theoretical density. However, BW32 is shown to be lower than its theoretical density. This might be due to the way the boron and tungsten atoms are arranged inside the PE matrix, reducing the amount of matter per cubic cm of the composite as illustrated in Figure 3.

$$\rho_C = \rho_{PE} w_{PE} + \rho_B w_B + \rho_W w_W \tag{1}$$

$$\rho_C = \frac{1}{\left(\frac{w_{PE}}{\rho_{PE}} + \frac{w_B}{\rho_B} + \frac{w_W}{\rho_W}\right)} \tag{2}$$

where ρ_C density of composite (g/cm^3)
ρ_{PE} density of polyethylene (0.86 g/cm^3)

ρ_B density of boron (2.34 g/cm^3)
ρ_W density of tungsten (19.25 g/cm^3)
w_{PE} weight fractions of polyethylene
w_B weight fractions of boron
w_W weight fractions of tungsten

Table 2. Densities of samples

	Theoretical density (g/cm^3)		*Measured density*
Sample	*Lower bound*	*Upper bound*	*(g/cm^3)*
PE100	–	–	0.86
B25	1.02	1.23	1.03
W25	1.13	5.46	1.16
BW32	1.15	4.04	1.00

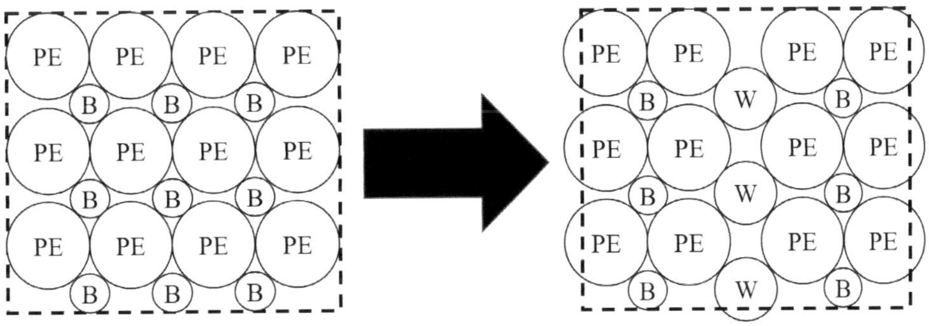

Figure 3. When tungsten atoms are added into the matrix of polyboron (a), the amount of matter in the same given volume decreases as shown in (b). Note that this is an oversimplification since polyethylene molecule is a long chain of $(C_2H_4)_n$ monomers.

3.2. Shielding performance of samples

Figure 4 shows the transmission ratio for all samples. In terms of neutron shielding, B25 is the best which might be due to its high boron content. However, at 5 cm, there is an increase in the neutron dose rate from B25. This might be an effect of buildup in which the neutrons are reflected by the shield atoms towards the detector. The second best is the pure polyethylene PE100. This might be because it has the highest hydrogen content among all samples which are very effective in scattering neutrons. This is also the reason why the new composite BW32 is the worst neutron shield. It has the lowest hydrogen content which affected its capability in slowing down the neutrons. Therefore, its resulting neutron dose rates are higher than the other samples. Nevertheless, all samples show a decrease in their transmission ratio when their thickness increases.

Table 3 shows the comparison of the neutron shielding performance between the composites in this study and previous studies. The removal cross-sections Σ_R are calculated using Equation 3, while the mass removal cross-sections are obtained by dividing Σ_R by density. B25 has the highest mass removal cross-section than the past studies due to having the highest boron content of 25 % by weight. W25 is expectedly worse than the past studies due to having no boron content. This is also the case with the new BW32. Although it has more boron than the past studies, it has less hydrogen content due to the additional content of tungsten. This lessens the effect of neutron moderation, resulting in higher neutron dose rates. Nevertheless, this new composite has good mixed radiation shielding performance as discussed later.

$$\Sigma_R = \frac{1}{x}\ln\frac{D_0}{D} \tag{3}$$

where
Σ_R fast neutron removal cross-section (cm^{-1})
x shield thickness (cm)
D_0 initial neutron dose rates ($mSv\ hr^{-1}$)
D final neutron dose rates ($mSv\ hr^{-1}$)

Table 3. Comparison of composite neutron shielding performances

Author	**Composition**	**B (wt%)**	**ρ (g/cm³)**	**Σ_R (cm^{-1})**	**Σ_R/ρ (cm^2/g)**
Sazali et al. (current)	PE100	0	0.86	0.1069	0.1243
	B25	25	1.03	0.1414	0.1373
	W25	0	1.16	0.0865	0.0746
	BW32	16	1.00	0.0520	0.0520
Uddin et al. (2020) [2]	HDPE+B	10	1.03	0.0981	0.0948
Abd Elwahab et al. (2019) [9]	HDPE+Borax	7.72	2.08	0.1570	0.0754
El-Khayatt (2010) [10]	PE+B	8.97	1.60	0.0953	0.0596

In Figure 4(b), all samples exhibit a decrease in gamma dose rates as their thickness increases, except W25 which shows a slight increase in transmission from 1 cm to 3 cm. This might be due to the inelastic scattering of neutrons and neutron capture by tungsten atoms (E_γ = 0.13, 0.48, and 0.69 MeV). It is expected to be the best gamma shield due to it having the highest tungsten additive. However, this is only observed at 1 cm thickness. At greater thicknesses, it is outperformed by B25 and BW32. B25 shows a steady decline in gamma-ray transmission. BW32 has the lowest gamma dose rates at 2-4 cm except at 1 cm and 5 cm. These gamma shielding performances might have been affected by the secondary gamma-rays generated from the scattering of neutrons and neutron capture by hydrogen (E_γ = 2.2 MeV), boron (E_γ = 0.48 MeV), and tungsten. The worst gamma shield was PE100. Even though it has good neutron attenuation as seen in Figure 4(a), it is not able to absorb the gamma-rays effectively without any additives.

Finally, in terms of mixed radiation attenuation, the new BW32 is the best among all samples as shown in Figure 5. Although it does not have the best neutron attenuation (Figure 4(a)), its combined neutron and gamma-ray transmission ratio is the lowest. This proves the effectiveness of combining various elements to improve the base material. In Figure 5, polyethylene by itself has poor mixed radiation performance as compared to others. By adding boron and tungsten, the transmitted radiations of both neutrons and gamma-rays can be lowered.

IOP Conf. Series: Materials Science and Engineering 1231 (2022) 012010 doi:10.1088/1757-899X/1231/1/012010

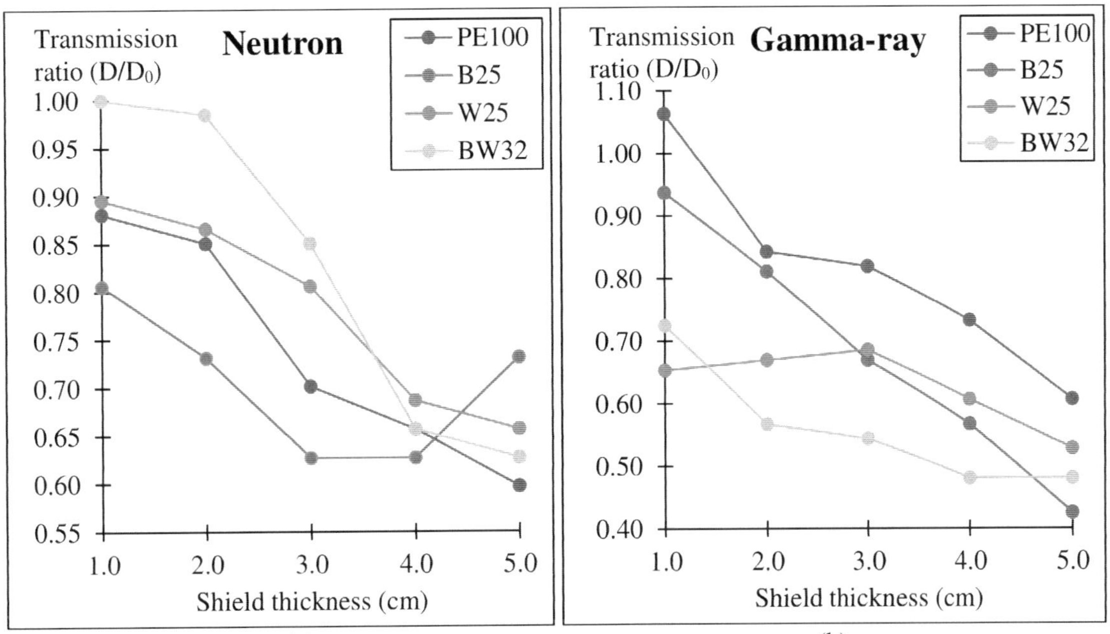

(a) (b)

Figure 4. Transmission ratio per shield thickness for neutron (a) and gamma-rays (b).

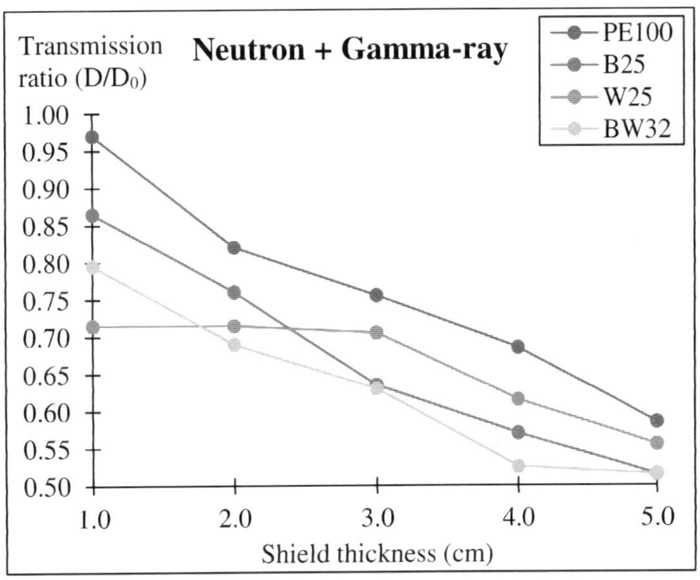

Figure 5. Total dose rate per shield thickness

4. Conclusion

Four samples of shielding are fabricated and tested through an attenuation experiment with a mixed neutron and gamma-ray source, ^{252}Cf. The first sample PE100 is pure polyethylene and it is used as the control to observe the effect of the additives in the composite samples. PE100 shows good neutron attenuation but has the worst gamma-ray attenuation. The second sample is B25, a polyethylene-based composite with 25% by mass of boron. It has the best neutron attenuation but at 5 cm, the transmitted dose rate increases due to the buildup effect. Its gamma transmission decreases almost linearly with thickness. The third sample W25 is also a composite with a PE base and tungsten powder (25% by mass). It is slightly worse than PE100 in terms of neutron shielding due to lower hydrogen content. Contrary to expectation, W25 does not exhibit the best gamma attenuation, and the gamma dose rates even increase slightly at thicknesses of 1-3 cm probably due to inelastic scattering of neutrons. Finally, BW32 shows the best mixed neutron and gamma-ray attenuation, proving the advantage of combining various elements to improve the shielding capabilities of the base material. Comparisons to previous studies are also provided. Future researchers may expand this study by investigating and improving its structural strength or other relevant properties. The methodology can be improved by using radiation sources of higher activities and using narrow beam geometry to identify the buildup factor.

References

[1] More C V, Alsayed Z, Badawi M S, Thabet A A and Pawar P P 2021 Polymeric composite materials for radiation shielding: a review *Env. Che. Let.* 1-34.

[2] Uddin Z, Yasin T, Shafiq M, Raza A and Zahur A 2020 On the physical, chemical, and neutron shielding properties of polyethylene/boron carbide composites *Rad. Phy. & Chem.* **166** 108450

[3] Nurazila M Z, Hafizal Y and Harun M M 2020 Neutron attenuation and mechanical properties of polymer composites filled with boron carbide particles *IOP Conf. Ser.: Mat. Sci. & Eng.* **785(1)** 012005

[4] Jumpee C, Rattanaplome T and Kumwang N 2020 A wide energy range neutron shielding material based on natural rubber and boron *IOP Conf. Ser.: Mat. Sci. & Eng.* **773(1)** 012036

[5] Alavian H and Tavakoli-Anbaran H 2019 Study on gamma shielding polymer composites reinforced with different sizes and proportions of tungsten particles using MCNP code *Prog. in Nuc. Ener.* **115** 91-98

[6] Asgari M, Afarideh H, Ghafoorifard H and Amirabadi E A 2021 Comparison of Nano/Micro Lead, Bismuth and Tungsten on the Gamma Shielding Properties of the Flexible Composites against Photon in Wide Energy Range (40 keV to 662 keV) *Nuc. Eng. & Tech.* **53(12)** 4142-4149

[7] Can Ö, Belgin E E and Aycik G A 2021 Effect of different tungsten compound reinforcements on the electromagnetic radiation shielding properties of neopentyl glycol polyester *Nuc. Eng. & Tech.* **53(5)** 1642-1651

[8] Callister W D, Rethwisch D G 2010 *Materials Science and Engineering: an Introduction, Materials Science and Engineering an Introduction*

[9] Abd Elwahab N R, Helal N, Mohamed T, Shahin F, Ali FM 2019 New shielding composite paste for mixed fields of fast neutrons and gamma rays. Mater. Chem. Phys. **233** 249–253

[10] El-Khayatt A M 2010 Calculation of fast neutron removal cross-sections for some compounds and materials *Ann. Nucl. Energy* **37** 218–222

Acknowledgement

We would like to thank Nuclear Malaysia and Universiti Teknologi Malaysia for providing the necessary facility and equipment for the completion of the project.

IOP Conf. Series: Materials Science and Engineering 1231 (2022) 012011 doi:10.1088/1757-899X/1231/1/012011

Correlation of ground penetrating radar and nuclear density gauge data to determine the density of asphalt pavement

Tengku Sarah Tengku Amran[1,a)], Mohamad Syafiq Mohd Amin[1], Mohamad Ridzuan Ahmad[1], Noor Azreen Masenwat[1], Suhairy Sani[1], Mohd Fajri Osman[1], Shaharudin Sayuti[1], Syed Abdullah Hamidi Syed Mamat[2], Haziq Hassan[1] and Anis Iwana Rohesan[3]

[1]Non Destructive Testing-Material for Structure Integrity (NDT-MSI), Industrial Technology Division, Malaysian Nuclear Agency (ANM) Bangi, 43000 Kajang, Malaysia.
[2]Faculty of Science and Technology, Universiti Kebangsaan Malaysia (UKM),43000 Bangi, Malaysia.
[3]School of Physics, Universiti Sains Malaysia (USM), 11800 USM, Penang, Malaysia.

[a]sarah@nm.gov.my

Abstract. This paper demonstrates the field performance of Ground Penetrating Radar (GPR) as a non-destructive testing tool to predict in situ asphalt pavement. In previous study, GPR has been used successfully in a variety of pavements application such as determining layer thickness, detecting subsurface distresses, and others. Nuclear Density Gauge (NDG) is specifically designed to measure the moisture and density of soils, aggregates, cement, and lime treated materials, and also to measure the density of asphalt concrete. Alternative analysis approaches by correlating the GPR and NDG data had been performed for density prediction of asphalt pavement. A density model name Al-Qadi, Lahouar, and Leng (ALL model) was used to predict the density of the asphalt pavement from the GPR data correlate with NDG database. This study shows that density for asphalt pavement from GPR data provide reasonably accurate density prediction when ALL model approach was used.

1. Introduction

Asphalt mixture density greatly influences the performance of asphalt concrete (AC) pavement. Appropriate AC density increases stability and durability, and reduces maintenance and rehabilitation costs of pavement. Insufficient density leads to pavement distresses including cracking, raveling, and/or oxidation [1]. Traditional density estimation methods are through a destructive method – conducting lab-controlled tests on cores from the field. This approach is time consuming and expensive.

Non-destructive testing (NDT) techniques have been developed by determining the structural condition of an in-service pavement. NDT has two major advantages over destructive testing. First, destructive testing by definition disturbs or requires removal of the pavement materials to a laboratory for testing, whereas NDT is truly an in situ test that evaluates the pavement without any material disturbance or modification [2]. The second advantage of NDT method is providing access to material properties while remaining rapid and inexpensive, allowing more tests to be completed while causing less disruption to traffic than destructive testing [3].

Meanwhile, electromagnetic (EM) wave methods have been successfully developed and implemented for AC pavement density prediction [4]. An alternative approach is to use nuclear density gauge (NDG) [5]. Although the nuclear gauge is a non-destructive technique that provides reasonably accurate estimates of the AC layer density, this technique, too, has some drawbacks. One of the drawbacks with NDG method is the use of radioactive material and thus requires special licensing to transport, which entails increased operational costs [6].

Researchers have tried to develop techniques to overcome these disadvantages. The investigation of using ground-penetrating radar (GPR) as an alternative to the nuclear density gauge to predict in situ asphalt mixture density continuously and rapidly [7,8]. GPR has the most potential because GPR surveys are nondestructive, rapid, continuous, and can provide multiple types of pavement structure information (e.g., layer thickness and substructure distresses) in addition to density [9]. GPR uses air-coupled antennas to enable high-speed (up to 96 km/h) and large-coverage (up to 2 m in transverse direction) survey. However, the GPR raw data, i.e., radargram, depends on the survey speed, sampling rate, and d properties. Therefore, the radargram is a distorted image of the subsurface condition, and specialized post-processing algorithms are needed for data interpretation [4].

To investigate the feasibility and effectiveness of using GPR to predict pavement density, several density prediction models were developed according electromagnetic mixing theory [8]. These density models enable the prediction of pavement density using GPR requires correlations between AC mixture dielectric constant and density [10]. This study uses the Al-Qadi, Lahouar, and Leng (ALL) density prediction model because its accuracy has been identified as best performance model to estimate pavement density [9,10].

The focus of this study is to develop correlation of GPR and NDG data in which to determine the density of asphalt pavement while increased capabilities of GPR as complete system for local road applications with using the ALL model equation. The GPR data used in this project was collected from the asphalt pavement test site. The accuracy of these predicted densities was validated from GPR data and compared with densities measured by the nuclear density gauge. The alternative analysis approaches that could be applied to improve the GPR accuracy when the optimum conditions are not met are also discussed.

2. Experimental Works

2.1. Experiment setup

This experiment was held on asphalt pavement in Block 60, Malaysia Nuclear Agency. The dimension of the asphalt scanning area was 2 m × 2 m with 9 points marked on the surface of the asphalt. The distance between each point were 1 m diagonally and 1 m horizontally. The GPR used was GSSI 900MHz shielded antenna and in air-coupled mode. The distance between shielded antenna and asphalt pavement was 25 cm and was held by GSSI cart system. Data was collected using SIR 4000 and then processed using RADAN 7 software. For comparison of the asphalt density value, NDG method was used to calculate the density of the asphalt using Troxler NDG system.

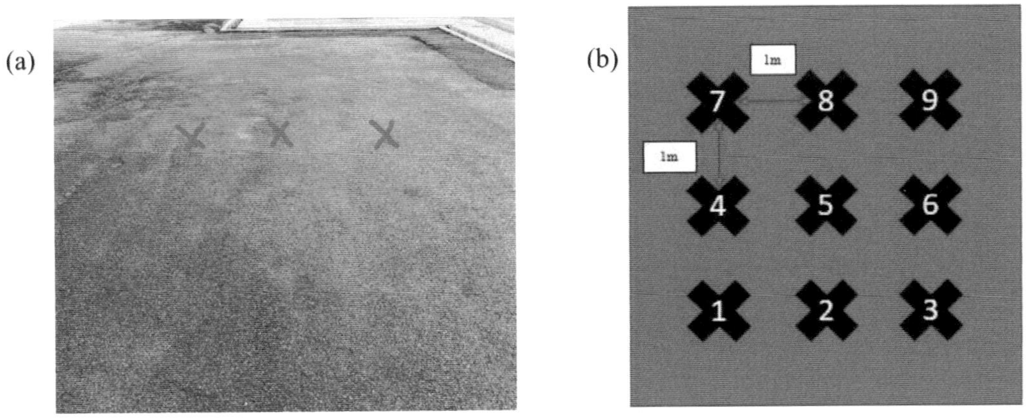

Figure 1. (a) Experiment set up on asphalt (b) Schematic diagram of the experimental set up.

Figure 2. a) The GPR GSSI 900 MHz equiptment on GPR cart. b) Troxler NDG equipment

The experiment was divided by two parts, named Experiment 1 and Experiment 2, where the difference are thickness of metal plate and the arrangement of metal plate on measured pavement. Experiment 1 was to compare the effect of metal plate thickness while Experiment 2 is to utilise the best performance metal plate in Experiment 1 and improving other technical setups.

(a) (b)

Figure 3. Metal plate with (a) 2mm and (b) 5mm thickness

During experiment 1 the metal plates were put in every point to find the average of A_{AC} meanwhile for experiment 2 the metal plate was put in the middle of the asphalt to find single point A_{AC} .Experiment 1 needed double scanning to obtain average A_{AC} for the first scanning and A_0 for the second scanning. Meanwhile for the experiment 2, it just needed single scanning in obtaining A_0 and A_{AC} for G_{mb} calculation of the asphalt.

(a) (b)

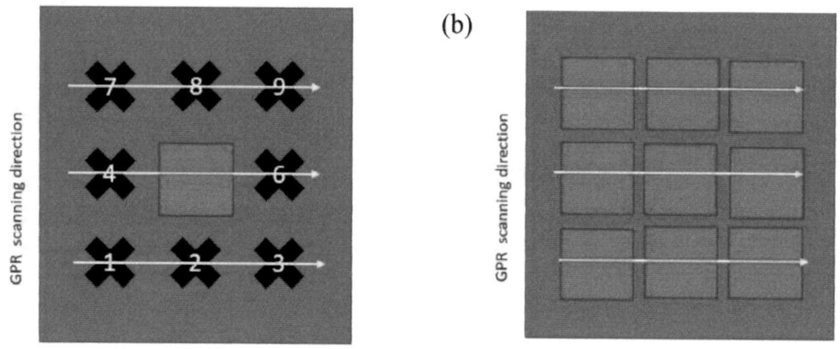

Figure 4. a) Schematic diagram of single point scanning. b) Schematic diagram for multi points scanning.

2.2. Calculation formula

The density of the asphalt can be calculated using ALL Model. This model used density constant from signal from GPR radargram that obtained ε_{AC}value from EM mixing theory. Then, ε_{AC} was used to back calculated the value of ε_s for the further calculation. Finally, the G_{mb} value were calculated for each point for every experiment. The formula for ε_{AC} (equation 1) and G_{mb} (equation 2) calculation is showed below:

$$\varepsilon_{AC} = \left(\frac{1+\left(\frac{A_0}{A_{AC}}\right)}{1-\left(\frac{A_0}{A_{AC}}\right)} \right)^2 \quad (1)$$

where ε_{AC} is dielectric constant of the top layer, A_0 is the amplitude of the surface reflection and A_{AC} is the amplitude of the incident GPR signal that produced by the plate. The G_{mb} and ε_s are obtained using ALL Model as shown below:

$$G_{mb} = \frac{\left(\frac{\varepsilon_{AC}-\varepsilon_b}{3\varepsilon_{AC}-2.3\varepsilon_b}\right)-\left(\frac{1-\varepsilon_b}{1-2.3\varepsilon_b+2\varepsilon_{AC}}\right)}{\left(\frac{\varepsilon_s-\varepsilon_b}{\varepsilon_s-2.3\varepsilon_b+2\varepsilon_{AC}}\right)\left(\frac{1-P_b}{G_{se}}\right)-\left(\frac{1-\varepsilon_b}{1-2.3\varepsilon_b+2\varepsilon_{AC}}\right)\left(\frac{1}{G_{mm}}\right)} \quad (2)$$

where G_{mm} maximum is specific gravity, G_{se} is effective specific gravity of aggregates, P_b is binder content that can be obtained from mix design. The value is set 0.05 for Pb, 2.641 for G_{se} and 2.521 for G_{mm}. ε_b is dielectric of binder that set as 3 [11]. ε_s is back calculated using the density that obtained from NDG result.

Root Mean Square Error (RMSE) also calculated for this experiment to compare the NDG and GPR data. RMSE is the square of all the error. It is excellent general-purpose error metric for numerical prediction according to P.Neil (2018)[12].

$$RMSE = \sqrt{\frac{1}{n}\sum_{i=1}^{n}(Si - Oi)^2} \quad (3)$$

where Oi is the NDG results and Si is G_{mb} predicted values for GPR.

3. Result and analysis

The data was obtained from radargram by using SIR4000 and then was processed by Radan 7 software. Twelve radargram was obtained for experiment 1 and three radargram was obtained for experiment 2. A_0 and A_{AC} were obtained from radargram using Radan 7 software that will be used to calculate value of ε_s. ε_s was calculated using ALL model for each experiment and G_{mb} was calculated using the value of ε_s .

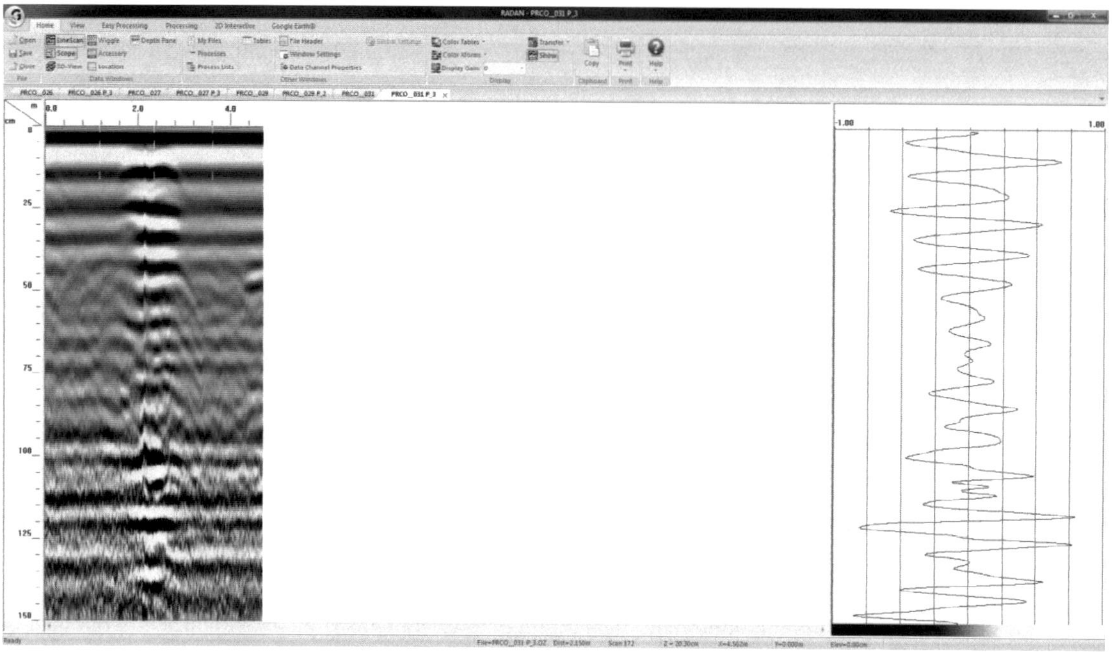

Figure 5. GPR GSSI 900 MHz radargram with scope.

3.1. Experiment 1

The value of asphalt density using NDG was shown in the table below. The value of average A_{AC} calculated for 2mm metal plate is 0.449192 and 0.471592 for 5mm metal plate. By using the value of average A_{AC} and value of A_0 radargram, ε_{AC} was calculated using density model for each metal plate. By using point 5 as reference point, ε_s was calculated using ALL model. ε_s for 2mm metal plate is 1.2703 meanwhile ε_s for 5mm metal plate is 1.3234.

Table 1. G_{mb} obtained from NDG reading.

G_{mb} Obtained from NDG									
Point	1	2	3	4	5	6	7	8	9
Gmb	1.893	1.822	1.897	1.908	1.902	1.835	1.896	1.894	1.867

Table 2. ε_{AC} calculated by using density model.

ε_{AC} Value for 2mm metal plate					
Point	1	2	3	4	5
Eac	1.583694	1.711280	1.646167	1.776755	1.776755
Point	6	7	8	9	
Eac	1.709078	1.776691	1.776691	1.709079	

ε_{AC} Value for 5mm metal plate					
Point	1	2	3	4	5
Eac	1.439569	1.57797	1.549207	1.565185	1.639137
Point	6	7	8	9	
Eac	1.481084	1.667686	1.520996	1.577970	

Thus, G_{mb} for each point was calculated and by using ALL model equation and percentage error for each point was calculated between GPR G_{mb} and NDG G_{mb}.

Table 3. G_{mb} estimation using ALL model equation for 2mm metal plate.

G_{mb} for GPR 2mm metal plate									
Point	1	2	3	4	5	6	7	8	9
Gmb	1.938	2.025	2.018	1.902	1.902	2.027	1.902	2.017	2.027
Error (%)	2.32	10.02	6.00	0.31	0.00	10.46	0.32	6.49	8.69

Table 4. G_{mb} estimation using ALL model equation for 5mm metal plate

G_{mb} for GPR 5mm metal plate									
Point	1	2	3	4	5	6	7	8	9
Gmb	1.474	1.810	1.752	1.786	1.902	1.588	1.925	1.688	1.810
Error (%)	22.13	0.66	7.64	6.39	0.00	13.46	1.53	10.88	2.95

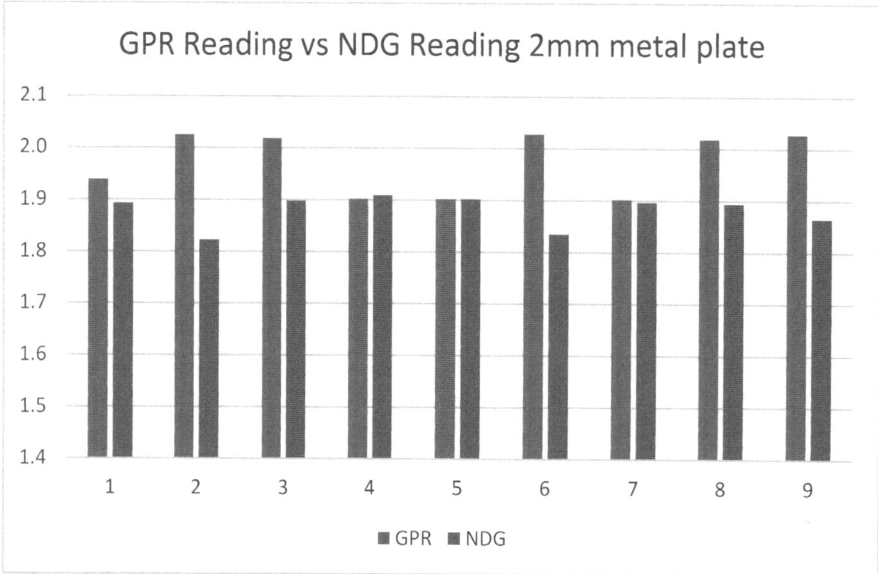

(a)

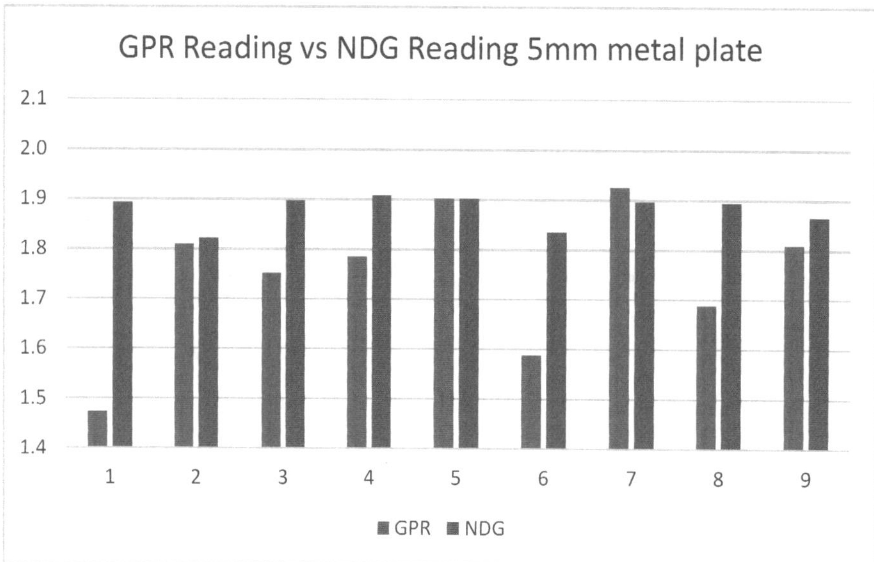

(b)

Figure 6. Difference G_{mb} value between GPR and NDG (a) 2mm metal plate (b) 5mm metal plate.

Based on experiment 1 result, it shows that 2mm metal plate gives more accurate G_{mb} results compared to G_{mb} value of 5mm metal plate. Besides, percentage error of 2mm metal plate was ranged from 0.32 to 10.46 percent of error with NDG result compared to 5mm metal plate which is ranging from 1.53 to 22.13 percent of error. This is also showed that GPR also capable in measuring density of the asphalt by an error of 10% compared to NDG result. Root Mean Square (RMS) error also calculated between G_{mb} result between GPR and NDG for each plate. For 2mm metal plate RMS error was 0.12301 meanwhile for 5mm metal plate was 0.188245.

3.2. Experiment 2

The values of asphalt density obtained from NDG and were converted into G_{mb} values are shown in the Table 5. For the Experiment 2, based on the results from the Experiment 1 which is thinner plate, 2mm, is used as it performs better in term of accuracy of density value in respect of the value from NDG. Besides, also from the Experiment 1, the data shows that the A_0 value is approximately same from each point. Hence, for practicality, the 2 mm metal plate is placed only on one point which is point 5. This useful when considering that this technique will used in on-field, which is on real roads, there is no need to place metal plate along the road. Therefore, A_{AC} value was obtained from one point, which is at point 5, resulting no need to calculate average value of A_{AC}. The A_{AC} value in the experiment 2 is 0.286325.

Table 5. G_{mb} obtained from NDG reading

	G_{mb} Obtained from NDG								
Point	1	2	3	4	5	6	7	8	9
Gmb	1.856	1.747	1.671	1.889	1.904	1.885	1.791	1.818	1.861

ε_{AC} value is calculated by using the ε_{AC} formula using the obtained A_0 and A_{AC} value from GPR radargram shown in the Radan 7 software. Table 6. shows the value of ε_{AC} calculated for each point.

Table 6. ε_{AC} calculated by using density model

	ε_{AC} Value								
Point	1	2	3	4	5	6	7	8	9
Eac	5.933575	4.478624	4.183906	5.522527	5.522527	5.993575	6.382303	3.911595	5.933575

Then, ε_s value is back-calculated from the ALL model formula by using G_{mb} value obtained from NDG and ε_{AC} value at point 5. The value of ε_s is 10.73686093.

Since the ε_s value is obtained, G_{mb} value from GPR is possible to calculated using the ALL model formula. Table 7. shows the G_{mb} value calculated from GPR and error percentage in respect of difference of the value from G_{mb} of NDG. The difference in G_{mb} value from NDG and GPR is illustrated in Figure 7.

Table 7. G_{mb} estimation using ALL model equation for 2mm metal plate

	G_{mb} Value from GPR								
Point	1	2	3	4	5	6	7	8	9
Gmb	1.947	1.845	1.856	1.903	1.903	1.947	2.002	1.886	1.947
Error (%)	4.90	5.61	11.07	0.74	0.05	3.29	11.78	3.74	2.95

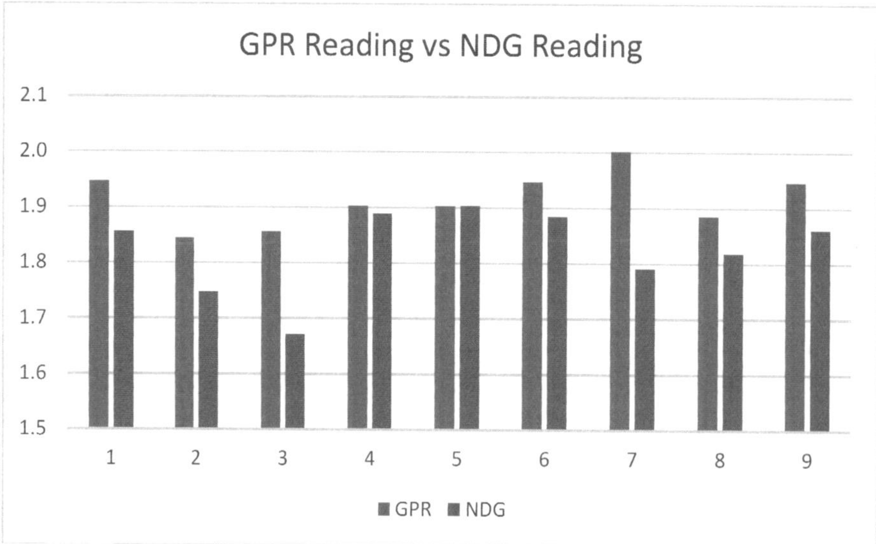

Figure 7. Difference G_{mb} value between GPR and NDG

Based on experiment 1 result, 2mm metal plate showed promising results in terms of its low percentage error in G_{mb} values compared to 5mm metal plate. That error percentage from 2mm metal plate ranging from 0.32% to 10.46% while 1.53% to 22.13% were achieved from 5mm metal plate experiment. This is due to thicker metal that causes radio waves from GPR to interfere and affect the radargram imaging resulting inaccurate A_0 value and further affect the calculation. These data gained from experiment 1 leads to the idea of experiment 2 which was to improve G_{mb} value, which means to achieve lower error percentage. However, instead of using multiple metal plate, experiment 2 was done by using only one 2mm metal plate instead. This is due to practicality, where this method will be used on-site, it will be more difficult if every point needs a metal plate.

In experiment 2, point 3 and point 7, shows a quite large error percentage compared to others, which are 11.07% and 11.78% respectively. Noted that experiment 2 is done slightly after rainy weather resulting asphalt to contain water. This could raise a problem as water can affect the density of the asphalt. As for GPR, it may affect the radargram image, due to blurry image, and this gives difficulty during the process of obtaining the value of A_0 and A_{AC}. This problem could give error in obtaining those values and will affect further calculation. Root Mean Square (RMS) error also calculated between G_{mb} result between GPR and NDG for 2mm metal plate is 0.111899.

4. Conclusion

The objective of the study was to determine the correlation of ground penetrating radar and nuclear density gauge data to determine the density of asphalt pavement can be achieved by using appropriate model such as ALL model. As part of this task, two experiment was conducted to improve the quality of the data collected. These include a comparison of two metal plate in Experiment 1 and application of the best metal plate in Experiment 2.

Experiment 1 concluded that 2mm metal plate is more accurate than 5mm metal plate which result in the G_{mb}value closer to the G_{mb} value on NDG. The experiment using 2mm metal plate has error percentage ranged from 0.32% to 10.46% compared to 1.53% to 22.13% in 5mm metal plate experiment. To achieve more practicality and accuracy, experiment 2 was planned by using only one point of 2mm metal plate rather than metal plate on each point.

Experiment 2 has error ranged from 0.05% to 11.78% from nine points. Root Mean Square (RMS) error also calculated between G_{mb} result between GPR and NDG for 2mm metal plate is 0.111899. This concluded that GPR performance in density measurement of asphalt is impressive and comparable to NDG except for some points that have larger error percentage. Fortunately, for further experiment, these errors may be lowered if the experiment is done on a hot sunny day where the absence of water on asphalts and soil beneath it to achieve lower error percentage.

5. References

[1] Shangguan P, Al-Qadi I L , Leng Z, Schmitt R L, and Faheem A 2013 Innovative approach for asphalt pavement compaction monitoring with ground-penetrating radar *Transp. Res. Rec.* **2347** 79–87, , doi: 10.3141/2347-09.

[2] Terzi S, Karaşahin M, Gökova S, Tahta M, Morova N, and Uzun I 2013 Asphalt concrete stability estimation from non-destructive test methods with artificial neural networks *Neural Comput. Appl.***23**(3–4) 989–997 doi: 10.1007/s00521-012-1023-1.

[3] Breysse D 2012 Nondestructive evaluation of concrete strength: An historical review and a new perspective by combining NDT methods*Constr. Build. Mater.* **33** 139–163 doi: 10.1016/j.conbuildmat.2011.12.103.

[4] Wang S, Al-Qadi I L and Cao Q 2020 Factors Impacting monitoring asphalt pavement density by ground penetrating radar *NDT E Int.***115** 102296 doi: 10.1016/j.ndteint.2020.102296.

[5] Zhao S and Al-Qadi I L 2019 Algorithm development for real-time thin asphalt concrete overlay compaction monitoring using ground-penetrating radar *NDT E Int.* **104** 114–123 doi: 10.1016/j.ndteint.2019.04.008.

[6] Shangguan P, Al-Qadi I, Coenen A and Zhao S 2016Algorithm development for the application of ground-penetrating radar on asphalt pavement compaction monitoring *Int. J. Pavement Eng.* **17**(3) 189–200 doi: 10.1080/10298436.2014.973027.

[7] Popik M, Maser K, and Holzschuher C 2010 Using high-speed ground penetrating radar for evaluation of asphalt density measurements *TAC/ATC 2010 - 2010 Annu. Conf. Exhib. Transp. Assoc. Canada Adjust. to New Realities*

[8] Imad A L, Al-Qadi I, Leng Z 2011 In-Place Hot mix asphalt density estimation using ground penetrating radar and 3-D finite element analyses in-place hot mix asphalt density estimation using.

[9] Leng Z, Al-Qadi I L, and Lahouar S 2011 Development and validation for in situ asphalt mixture density prediction models *NDT E Int.* **44**(4) 369–375 doi: 10.1016/j.ndteint.2011.03.002.

[10] Leng Z, Al-Qadi I L, Shangguan P and Son S 2012 Field application of ground-penetrating radar for measurement of asphalt mixture density: Case study of Illinois route 72 overlay," *Transp. Res. Rec.***2304** 133–141 doi: 10.3141/2304-15.

[11] Büyüköztürk O, Taşdemir M A, Güneş O, and Akkaya Y 2012 *Nondestruc. testing of mater. and struct.: Proc. of NDTMS-2011, Istanbul, Turkey, May 15-18, 2011* (RILEM Bookseries*)* **6** 855–860 doi: 10.1007/978-94-007-0723-8.
[12] Neill M R, Simon, Hashemi 2018 *Fund.of Ocean Renew. Energy: Generating Electr. from the Sea.*

Acknowledgments
The authors would like to extent their appreciations and sincere thanks to Ministry of Science, Technology and Innovation (MOSTI), Universiti Kebangsaan Malaysia (UKM), Universiti Sains Malaysia (USM) and Agensi Nuklear Malaysia as well as the Non-Destructive Testing-Material and Structure Integrity Group (NDT-MSI) for the support and co-operation.

The effect of various electrospinning parameter and sol-gel concentration on morphology of silica and titania nanofibers

Umar Abdillah[1], Hafizal Yazid[2, a)], Sahrim Ahmad[1], Nurulizzati Makhtar[3], Siti Zaubidah[3], Ruey Shan Chen[1], Nur Hadifah Syifa[4]

[1]Department of Applied Physics, Faculty of Science and Technology, UKM, 43600 Bangi , Selangor, Malaysia
[2]Agensi Nuklear Malaysia, 43000 Kajang, Selangor, Malaysia
[3]Faculty of Applied Sciences, Universiti Teknologi MARA, Negeri Sembilan Branch, Kuala Pilah Campus, 72000 Kuala Pilah, Negeri Sembilan, Malaysia
[4]Malaysian Institute of Aviation Technology, University of Kuala Lumpur, 43900 Sepang, Selangor, Malaysia

[a]hafizal@nuclearmalaysia.gov.my

Abstract. Nano scale fibers of SiO_2 and TiO_2 with various compositions of viscosity and parameter have been prepared by electrospinning the precursor of tetraethyl orthosilicate (TEOS) and titanium (IV) isopropoxide (TiP). Different concentrations of polymer precursor polyvinylpyrrolidone (PVP) and applied voltages were studied. Distance tip to collector of 15 cm and flow rate of 0.003 mL/h were fixed for entire electrospinning process. The morphology of nanofibers was evaluated versus different solution concentrations as well as different applied voltages. Concentration focused in this study included 5 %wt, 10 %wt and 15 %wt of PVP along with 15 kV – 17.5 kV of voltages applied. The as-spun nanofibers were characterized by SEM. Results showed that the average fiber diameter decreases with decreasing of polyvinylpyrrolidone (PVP) solution concentration.

1. Introduction

Titanium dioxide or titania which an inorganic compound with a chemical formula of TiO_2 was laboratory created in late 1800s. This semi-metallic material is produced from ilmenite, rutile, and anatase. It is naturally occurring oxide of titanium which odourless, insoluble and absorbent. Titania is widely used material both as a binder to the fabrication of other products and as a material on its own. With excellent properties, titania is used widely in various applications related to environmental cleaning and protection such as in building painting [1]. Blending titania and paint together may lead to cost reduction of building operation due to its ability to reflect sun light (heat) [2]. Moreover, due to highly oxidizing, cheap and chemically stable, titania is a suited material to be use as photocatalyst to deteriorate air pollutants. In addition, titania nanomaterial is used widely in photovoltaic applications; nanocrystalline TiO_2 film is a core element in dye-sensitized solar cell (DSSC) [3]. Furthermore, titania nanotube has been reported being use as hydrogen sensor on account of its high sensitivity [4].

Silicon dioxide which also called as silica is the most significant material of earth lithosphere with chemical formula of SiO_2. This inorganic compound was first documented as "TiO_2" by Swedish chemist in late 1824. This metalloid compound is colourless/white crystalline and can be found in many crystalline forms includes α-, β-quartz, α-, β-cristobalite and α-, β-tridymite [5]. Variety applications of silica nanomaterial have been reported nowadays due to it interesting properties. Owing to high melting point, silica (sand) is widely used as core ingredient in manufacturing many engineering related equipment as well as sand casting [6]. Moreover, the correct amount of silica sand improved the strength of cement mixture which make silica as a useful binding component in concrete production [7]. In addition, some publication in nanomedicine describes the use of silica nanomaterial in emerging therapy where silica nanospheres reported being used as Magnetic Resonance Imaging (MRI) contrast agent [8, 9].

Electrospinning represents a simple process of producing nanofibers. It is a unique method to produces ultrafine fibers with a simple setup [10]. The basic apparatus of electrospinning includes a high voltage power supply, pump, syringe, and grounded collector. At the beginning of the electrospinning process, high voltage power is connected to the syringe before the sol-gel is injected. Due to the voltage applied, an electrical field is generated from syringe tip to the grounded collector. As the sol-gel is injected, an electrical repulsive force directs the fiber in form of jet onto the collector [11]. The fibers then calcined at high temperatures to remove the polymer precursor and the desired ceramic phase is obtained. Novel metal oxide (titania) and metalloid (silica) nanofibers have been successfully prepared by electrospinning process [12, 13]. Based on reported studies by Mehran et al. (2018) [14] and Amariei et al. (2017) [15] , it is clear that characteristics such as fiber diameter and fiber morphology are strongly depending on the solution concentration.

In the present work, formation of titanium dioxide nanofibers and silicon dioxide nanofibers using the electrospinning technique as well as the effect of solution concentration and apply voltages on nanofiber morphology is investigated.

2. Methods and materials

2.1. *Materials*

Titanium isopropoxide (TiP) and tetraethyl orthosilicate (TEOS) from Sigma Aldrich was used as the source of titania and silica precursor. For polymer precursor, high molecular weight polyvinylpyrrolidone (PVP) (Mw 1,300,000) was purchased from Sigma Aldrich alongside acetic acid, HCL and ethanol.

2.2. *Preparation of Sol-gel*

To simplify the condition for analysis, a typical procedure was selected based on previous study. Both silica and titania polymer precursor were prepared by the same procedure. Initially, three different polymer precursor concentrations (5 %, 10 % and 15 %) for both silica and titania were prepared by adding 5 g, 10 g, and 15 g of polyvinylpyrrolidone (PVP) into 10 ml ethanol followed by magnetic stirring for 2 hours at 50 °C with every solution held in a closed beaker. Silica precursor was prepared by mixing TEOS with 3 mL ethanol in a glass beaker followed by adding HCl/water solution to the TEOS/ethanol solution under stirring process. The solution was heated at 80 °C for 30 min and then cooled down to room temperature. Prior to electrospinning, solution was mixed with prepared polymer precursor and undergoes 1 hour stirring process. Titania precursor was prepared by mixing titanium isopropoxide (TiP), 3 mL of acetic acid and 3 mL of ethanol. After a few minutes, the solution then was mixed with prepared polymer precursor and undergoes 1 hour stirring process prior to electrospinning.

2.3. *Electrospinning process*

Prior to electrospinning process, each solution was loaded into a plastic syringe equipped with a 23-gauge needle. The needle was connected to a voltage supply which is capable of generating DC voltage as high as 25 kV. When concentration was varied, all other parameters were maintained as the following: feeding rate (0.03 mL/min) and electric field strength (17.3 kV). When applied voltage was varied, the polymer concentration was kept at 10 %. The distance between the injection orifice and the collector was kept at 15 cm for all experiments. The relative humidity was controlled between 50 % and 60 % at room temperature throughout the entire electrospinning

Table 1. Parameters of electrospinning process. When solution concentration and applied voltages were varied, all other parameters were maintained.

Parameter	***Fix***	***Vary***
Syringe	23-gauge needle	-
Feeding rate	0.03 mL/min	-
Distcane tip-collector	15 cm	-
Concentration	10 %	5 %, 10 %, and 15 %
Voltage	15 kV	10 kV, 15 Kv, and 17.5 kV

2.4. Characterization

Nanofibers were obtained using scanning electron microscopy (SEM) and Field Emission Scanning Electron Microscope (FESEM) and later morphological studies of fibres before thermal treatment were done employing these images.

3. Results and discussion

From the SEM micrographs as shown in Figure 1, it can be observed that beaded fibers formed at low concentration sol-gel of 5 % for both silica (a) and titania (b) electrospun mats. This phenomenon commonly reported when the concentration of the solution used is too low. This is due to the insufficient amount of sol-gel viscoelastic force to overcome the repulsive forces of charge thus resulted on polymer drops on the fibers. The identical phenomenon also reported by Suresh et al. (2005) [16], which they concluded that while polymer tend to have low resistance at low viscosity, surface tension will take place as a major influence on final morphology which may resulted on polymer drops instead of fibers due to high surface tension thus introduced bead formation on the produced fibers. The formation of bead on fiber would significantly reduce surface area hence affect the nanomaterial performance [17]. Beaded fibers were considered as "poor" quality fibers and the electrospinning parameters are often optimized to eliminate beads on fiber.

(a) (b)

Figure 1. Effect of polymer precursor (polyvinylpyrrolidone) concentration on fiber morphology of silica and titania before calcine at 20 μm (silica) and 5 μm (titania).Silica 5 % (a) and titania 5% (b).

However, beaded fibers are less likely to formed for more viscous solutions. Increase solution concentration will enhance viscoelastic performances of the polymer sol hence able to match the electrostatic force of electrospinning leading to formation of smooth fibers. When viscosity increased, concentration of polymer solution increased as well due to chain entanglements [18]. Beaded free fiber were obtained for both silica as shown in Figure 2 and titania in Figure 3 at 10 % and 15 % of polymer concentration. Concentrations of 10 % and 15 % polyvinylpyrrolidone of polymer precursor solutions over solvent proved to be the optimum range in order to produce bead-free fibers for both silica and titania nanofibrous mats.

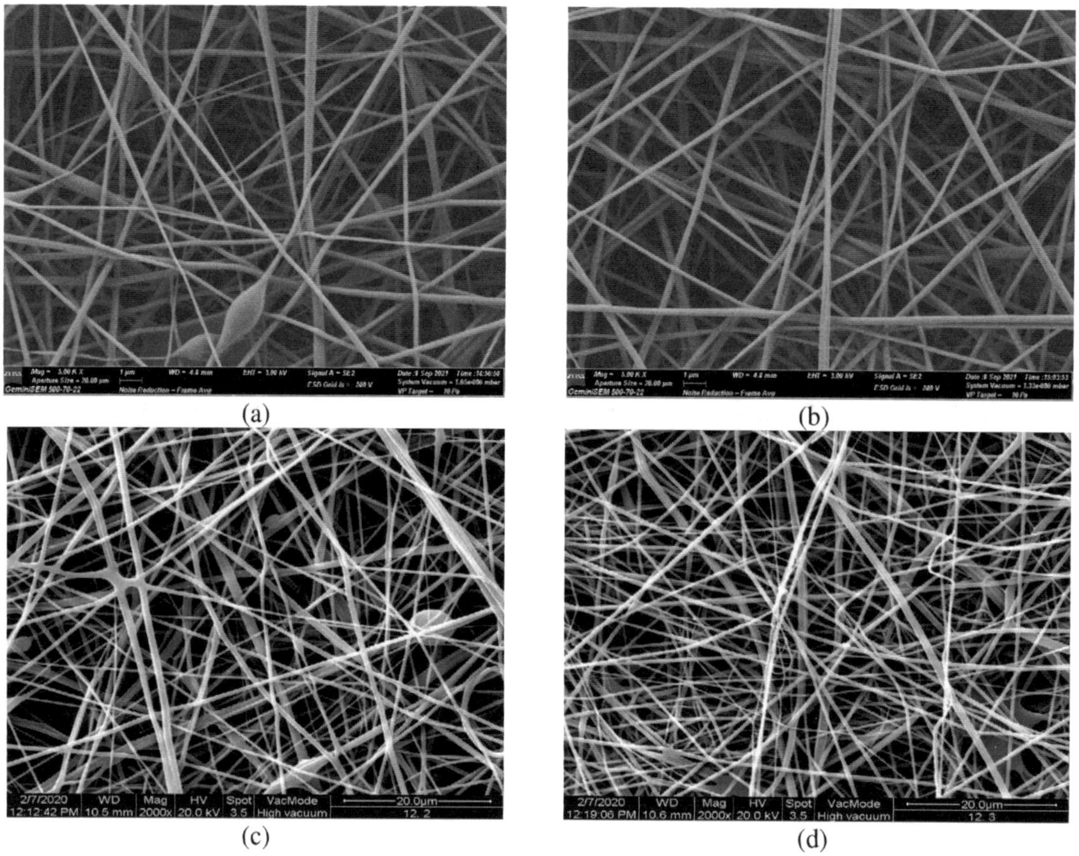

Figure 2. Effect of polymer precursor (polyvinylpyrrolidone) concentration on fiber morphology of silica and titania before calcine at 20 μm (silica) and 5 μm (titania). Silica fibers at 10 % (a) and 15 % (b) PVP concentration. Titania fibers at 10 % (c) and 15 % (d) PVP concentration.

The data in Figure 3 show trends for nanofibers diameter when PVP concentration become higher for both silica (a) as well as titania (b). 5 % polymer concentration of silica nanofiber (a) have an average fibers diameter of 110 nm. These diameter value then increased to 174 nm at 10 % polymer concentration and 210 nm at 15 % concentration. Titania (b) electrospun nanofiber also have identical phenomenon, where it can be observed that average fibers diameter of 5 % polymer concentration (125 nm) increased to an average of 166 nm (10 %) and 253 nm (15 %). The increments of polymer concentration would lead to several affect includes increases of polymer chain entanglements [16]. This phenomenon leading to thicker injected polymer from syringe thus increase the collected fiber diameter on grounded collector surface. Based on the literatures, removal of polyvinyl alcohol by calcination process of 500 °C will reduce the fiber diameter by an average of 80 nm [13, 19]. From this statement, an assumption was made and represented by the solid line (below) in Figure 3 for both silica (a) and titania (b). It can be observed that polymer concentration have significant effect on pure silica and titania

IOP Conf. Series: Materials Science and Engineering 1231 (2022) 012012 doi:10.1088/1757-899X/1231/1/012012

fiber diameter after calcination process. Increased in polymer concentration not only will increased fiber diameter before calcination but as well as pure fibers after removal of polymer precursor. When concentration is varied with fixed parameters in electrospinning, the volume of entrapped fiber precursor depends on volume of injected polymer, signifying that as the polymer become more concentrated it will resulted on the bigger fiber diameters [20].

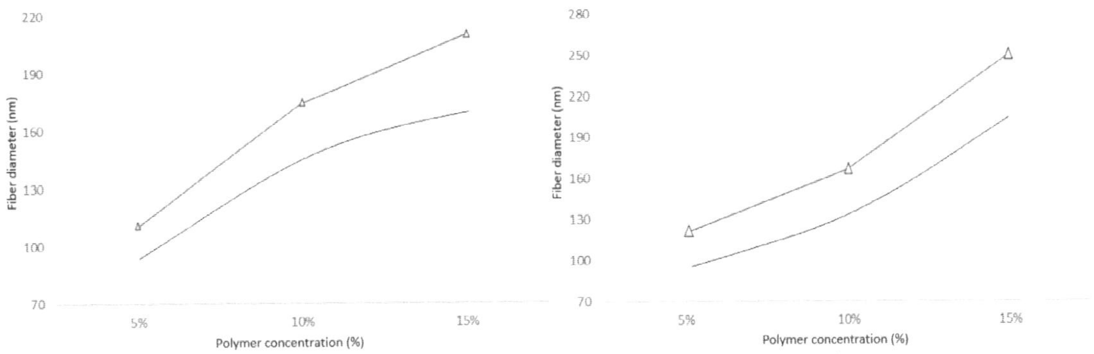

(a) (b)

Figure 3. Effect of PVP solution concentration for silica (a) and titania (b) on fiber diameter before calcine. Solid line (below) represents an assumption of calcined fiber diameter.

Adjusting the voltage applied have numbers of impacts on the electrospinning process. Increasing voltage will accelerate the electrospinning jet and this may result in larger volume of solution drawn from the tip of the needle. One of the impacts found resulted on increasing the applied voltage is reduction of fibers diameters. In this study, increasing the apply voltage resulted on the non-uniform fiber diameter distribution between 120-160 nm. These non-uniform fiber distribution was reported as the result of branching jet [21]. Accordingly, the distribution of non-uniform diameter of fibers in this study may resulted from branching jet during electrospinning process. However, fiber breakage is the most significant impact resulted from varying the applied voltage. As shown in Figure 4, fibers obtained at 10 kV appear broken whereas the fibers obtained at 15 kV and 17.5 kV are continuous. It can be concluded that applied voltage of 10 kV was not enough to produce continuous nanofibers from the solution at the given concentration.

(a) (b) (c)

Figure 4. Effect of voltage applied on fibers morphology with polymer concentration at 10 % and fixed other electrospinning parameters. (a) 10 kV, (b) 15 kV, (c) 17.5 kV.

4. Conclusion

In this work, the effect of PVP solution concentration on electrospun silica and titania fibers in terms of the morphologies has been explored. It was found that the properties of electrospun silica and titania fibers fabricated is critically depending on the electrospinning parameters such as PVP solution concentration. With decreasing the PVP solution concentration, the fiber diameter decreases. Beaded fiber structure was obtained at lower polymer concentration 5 % but an increase in the polymer concentration to 10 % and 15 % yielded bead free fibers, which indicates that a high viscosity is required to obtain uniform Silica and Titania nanofibrous mats. In addition, voltages apply have significant impact on electrospun fibers. Increasing the voltage will decrease the fiber diameter between 120-160 nm. The non-uniform fibers distribution may result from branching jet during electrospinning process. In addition, applied voltage of 10 kV was not enough to produce continuous nanofibers from the solution at the given concentration.

5. References

[1] Xiaobo Chen and Samuel S. Mao 2007 Titanium dioxide nanomaterials: synthesis, properties, modifications, and applications *Chem. Rev.* **107(7)** 2891–2959.

[2] Shuang Shi, Dongya Shen, Tao Xu and Yuqing Zhang 2018 Thermal, optical, interfacial and mechanical properties of titanium dioxide/shape memory polyurethane nanocomposites *Composites Sci. and Technol.* **164** 17-23.

[3] Boschloo Gerrit 2019 Improving the performance of dye-sensitized solar cells *Front. Chem.* **7** 77.

[4] Bertuna A, Comini E, Poli N, Zappa D and Sberveglieri G 2016 Titanium dioxide nanostructures chemical sensor *Procedia Engineering* **168** 313-16.

[5] Huahai Mao, Sundman B, Zhongwu Wang and Saxena S K 2001 Volumetric properties and phase relations of silica - thermodynamic assessment *J. of Alloys and Compounds* **327(1-2)** 253-62.

[6] Potapov V, Serdan A, Gorev D, Zubaha S and Shunina E 2019 Colloid silica in hydrothermal heat carrier: characteristics, technology of extraction, industrial applications *IOP Conf. Ser.: Earth Environ. Sci.* **249** 012043.

[7] Esfandiari J and Loghmani P 2019 Effect of perlite powder and silica fume on the compressive strength and microstructural characterization of self-compacting concrete with lime-cement binder *Measurement* **147** 106846.

[8] Kathryn M. L. Taylor, Jason S. Kim, William J. Rieter, Hongyu An, Weili Lin and Wenbin Lin 2008 Mesoporous silica nanospheres as highly efficient MRI contrast agents *J. Am. Chem. Soc.* **130(7)** 2154–55.

[9] Yan Chen, Kelong Ai, Jianhua Liu, Guoying Sun, Qi Yin and Lehui Lu 2015 Multifunctional envelope-type mesoporous silica nanoparticles for pH-responsive drug delivery and magnetic resonance imaging *Biomaterials* **60** 111-120.

[10] Nandana Bhardwaj and Subhas C. Kundu 2010 Electrospinning: A fascinating fiber fabrication technique *Biotechnol. Adv.* **28(3)** 325-47.

[11] Gregory C. Rutledge and Sergey V. Fridrikh 2007 Formation of fibers by electrospinning *Adv. Drug. Deliv. Rev.* **59(14)** 1384-91.

[12] Alves A K, Berutti F A, Clemens F J, Graule T and Bergmann C P 2009 Photocatalytic activity of titania fibers obtained by electrospinning *Mater. Res. Bulletin* **44(2)** 312-17.

[13] Li Zhang, Qinghong Zhang, Hongyong Xie, Jiang Guo, Hailong Lyu, Yaogang Li, Zhiguo Sun, Hongzhi Wang and Zhanhu Guo 2017 Electrospun titania nanofibers segregated by graphene oxide for improved visible light photocatalysis *Appl. Cataly. B: Environ.* **201** 470-8.

[14] Mehran Shahhosseininia, Saeed Bazgir and Morteza Daliri Joupari 2018 Fabrication and investigation of silica nanofibers via electrospinning *Mater. Sci. and Eng. C* **91** 502-11.

[15] Amariei N, Manea L R, Bertea A P,Bertea A and Popa A 2017 The influence of polymer solution on the properties of electrospun 3D nanostructures *IOP Conf. Ser.: Mat. Sci. and Eng.* **209** 012092

[16] Suresh L. Shenoya,Wm Douglas Batesa, Harry L. Frischb and Gary E. Wneka 2005 Role of chain entanglements on fiber formation during electrospinning of polymer solutions: Good solvent, non-specific polymer-polymer interaction limit *Polymer* **46(10)** 3372-84
[17] Adnan Haider, Sajjad Haider and Inn-Kyu Kang 2018 A comprehensive review summarizing the effect of electrospinning parameters and potential applications of nanofibers in biomedical and biotechnology *Arabian J. of Chem.* **11(8)** 1165-88.
[18] Jason Swei and Jan B Talbot 2003 Viscosity correlation for aqueous polyvinylpyrrolidone (PVP) solutions *J. of Appl. Polymer Sci.* **90(4)** 1153-55.
[19] Susan V Olesik and Toni E. Newsome 2014 Electrospinning silica/polyvinylpyrrolidone composite nanofibers *J. of Appl. Polymer Sci.* **131(21)** 40966.
[20] Hamid Esfahani, Rajan Jose and Seeram Ramakrishna 2017 Electrospun ceramic nanofiber mats today: Synthesis, properties, and applications *Materials* **10(11)** 1238.
[21] Yarin A L Kataphinan W and Reneker D H 2005 Branching in electrospinning of nanofibers *J. of Appl. Phys.* **98** 064501.

Acknowledgement
The authors would like to thank the Nuclear Malaysia and Ministry of Higher Education (MOHE) for the research financial support under the research scheme of Fundamental Research Grant (FRGS) with the grant number FRGS/1/2018/STG07/MOSTI/02/01.

Fabrication of magnetic sugarcane bagasse paper

Annur Azlin Azmi[1] and Siti Amira Othman[1, (a)]

[1]Department of Physics and Chemistry, Faculty of Applied Sciences and Technology, Universiti Tun Hussein Onn Malaysia, 84600, Pagoh, Johor, Malaysia

[a]sitiamira@uthm.edu.my

Abstract. Generally, development in the pulp and paper industry requires extensive cutting of trees, which in effect contributes to deforestation. The dramatic growth in demand for wood supply, combined with the increasingly increasing cost of timber, has created a surge of interest in the use of non-wood plant fibres for paper production in widely developed countries. The use of waste material in pulping and paper-based industries could be beneficial as it helps prevent the need for disposal, which currently increases agricultural costs and causes environmental deterioration due to pollution and fires. In this research, the sugarcane bagasse was dry and chopped into 5 cm in lengths. The fibre and pulp were separated and put in a pulp disintegrator then it was weighed and mixed with water. The paper that fabricates will be irradiated with gamma-ray and then followed by characterizing with Scanning Electron Microscope- Energy Dispersive X-Ray Analysis (SEM-EDX) and Fourier-transform infrared spectroscopy (FTIR). The SEM-EDX result shows that the weight and atomic percentage of the ferrite (Fe) increase after the irradiation. While for the FTIR, the entire sample exhibit the absorption range of 3400 to 2400 cm^{-1} region. The paper will be compared with the properties of multipurpose paper.

1. Introduction

Bagasse is the remaining residue of the fibrous after the crash of the sugarcane to extract its juice. Sugarcane bagasse among the most essential raw materials for paper and pulp manufacturing in other regions. The benefit of reducing deforestation will be introduced by using crop residues instead of wood. As bagasse is produced of non-wood fibre, therefore, it can be used as a substitute in replacing non-wood fibre as a sustainable source.

Sugarcane bagasse is an abundant lignocellulosic waste usually cultivated in open sugarcane processing countries. The sugar cane stalk includes two parts, the inner pit usually contains utmost sucrose and the outer rind containing lignocellulosic fibres. Throughout sugar processing, the sugar cane stalk is shredded in the process of extracting sucrose. This method generates a significant quantity of sugarcane bagasse residue which includes both pith fibres and crushed rind [1].

Besides that, the sugarcane residue after sugar cane extraction is one of the most widely available lignocellulosic fibre papermaking resources in certain developing countries. Approximately 54 million tons of bagasse are generated annually. These crop residues provide a plentiful, affordable, and widely accessible supply of sustainable lignocellulosic biomass. The chemical compositions of pure bagasse fibres bundles are cellulose, lignin, hemicelluloses, ash, and ethanol/dichloromethane extract.

Next, the addition of other material to paper processing became very necessary to significantly raise the efficiency of paper output. The filler usage in final chemistry and paper preservation help needs a certain degree of comprehension in the area of paper output chemistry. It can contribute to improved paper composition and, if used throughout the paper production process, maintain as much as the filler. For more than two decades, a filler is being used as an addition to enhance the consistency of material, such as clarity, stiffness, smoothness, and durability, also the manufacturing costs [2].

Bacteria and fungi may reduce or weaken the feature of the fibre in the sugarcane bagasse, which may lead to a significant presence of a spot or a stain in the paper. Simultaneously, the credibility of the paper will be affected since it is harmful to be used [3].

In a previous study, sugarcane bagasse has been exposed to gamma irradiation in various doses to know the changes in the characteristics for chemical, structural, physical, morphological, and properties of that sample itself. Demand to produce an eco-friendly, and cost-effective, the addition exposure was found effective in strengthening the digestibility of the lignocellulosic of the sugarcane bagasse [4].

It is believed that the non-wood option has a fibre quality equal to that of traditional wood products. This research offers useful knowledge on the implementation of non-wood raw materials to be used in the industry of paper in Malaysia. Besides, those non-wood fibres have a wide range of exceptional physical and optical features that might be leveraged to improve the quality of the product. As a result, it will ensure the sustainable utilization of natural resources.

2. Methodology

Sugarcane bagasse was sorting from dust and sand, by crushing, washed, filtering and air-dried to approximately 15% moisture content conducted prior to pulping. The pith was removed by around 30 percent. The sugarcane bagasse was dry depithed, crushed and chopped to about 5 cm length. The sugarcane bagasse fibre was stored in an oven for 2 hours After the pulp was separated from the fibre. It was transferred for the drying process. The process took about 5 to 10 minutes to ensure the pulp is dry before transferring it. The sugarcane bagasse pulp was undergoing a blended process until it becomes fine pulp about 5 minutes. After that, the pulp was crushed in the machine until it becomes a small lump of pulp. Then the pulps and ferrite magnet were weighed 20g and 7.5g each samples respectively using weighing balance before mixed with standard pulp disintegrator. The magnetics sugarcane bagasse pulp was blended with 2 liters of water in a Standard pulp disintegrator Model UEC-2008. In aqueous suspension, a known amount of pulp is drained through the screen plate of standardized slits to detach the fibre and the degradation is dependent on the disparity in size between them. This process is to know the smoothness of flow and disintegration of the pulp. A 12 liters of water was added to the solution of pulp in the vessel when the compressor was on. This machine of the drainage system generates a consistent flow through the whole wire, thereby making extremely uniform sheets. After the drainage process, the pulp that was embedded with the filler which is a ferrite magnet then was collected for about 1 liter.

The paper was created when the water was pumped out and leave the paper on top of it. Magnetics sugarcane paper was compressed with a rapid sheet drying machine. This method is used to reduce the drying time of wet pulp or paper samples. Flat heating elements are required for it. Magnetics sugarcane bagasse was prepared. The paper was cut into 5cm × 5cm to radiate in gamma cell irradiator. This is because, the lower the surface area, the higher the radiation can focus on the magnetic paper and the ionizing radiation particle. The properties, surface, structure of the magnetic paper was observed pre- and post-radiation to identify and observe the changes in the paper.

3. Results and discussion

3.1 *Characterization with Scanning Electron Microscope- Energy Dispersive X-Ray Analysis (SEM-EDX)*

The morphology of irradiated sugarcane bagasse as shown in Figure 1(a) clearly showed the shape of the microcrystalline cellulose structure of the bagasse. These shapes of cellulose in the lignocellulosic structure have appeared. The presence of these shapes of cellulose in the sugarcane bagasse paper indicate the complex structure in the components of it. Figures 1(a) shows the SEM images of the irradiated magnetics sugarcane bagasse paper. On the surface of it, exists as there is a net morphological, perhaps due to reaction between chemicals used. EDX analysis showed low presences of Fe in this fraction Figure 1(b). EDX spectrum Figure 1(b) also shows high Fe content which is 5.96% weight and 1.52% the atomic percentage of the Fe.

(a) (b)
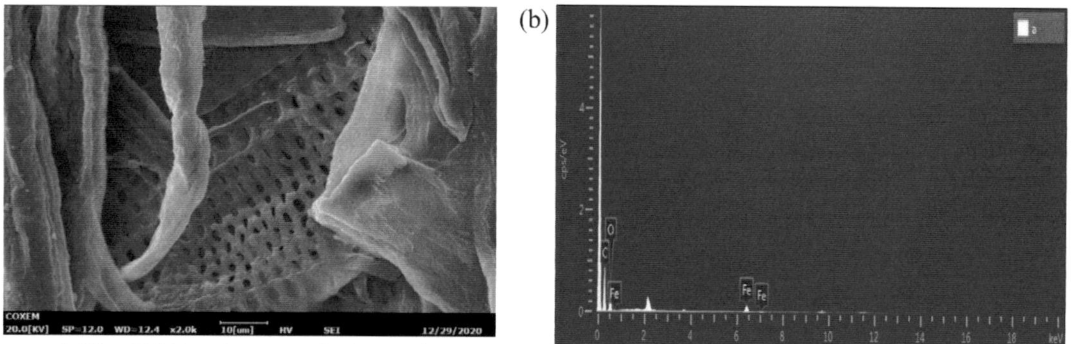

Figure 1. The SEM surface morphology micrograph of sugarcane bagasse fibre (a) surface morphology, (b) The EDX pattern of bonding between the magnetic particle in the lumen of the fibre for soda process technique.

Figure 2(a) shows the micrograph of multipurpose paper. From the image (white round shape) it can be seen that the fibre has uniform, smooth, and homogenous surface. EDX analysis showed no presences of Fe in this fraction as there is no ferrite used in this particular sample. EDX spectrum Figure 2(b) also shows no presence of Fe content in weight and the atomic percentage. This is due to no element of ferrite magnet that is associated with this multipurpose paper.

(a) (b)
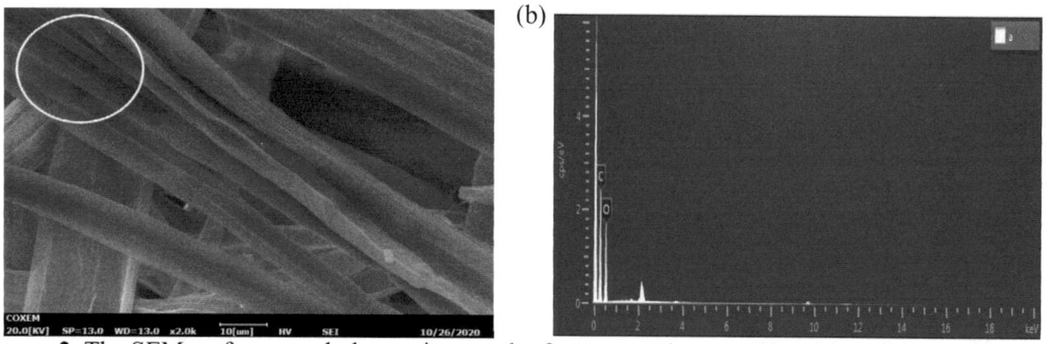

Figure 2. The SEM surface morphology micrograph of sugarcane bagasse fibre (a) surface morphology (b) The EDX pattern of bonding between the particle presents in the multipurpose paper.

Figure 3(a) represents the compact structure of sugarcane bagasse fibre. The micrograph shows the deconstruction of the compact rigid structure into cellulose fibres that appeared in irregular uniform strips reveals as a loosen bundles. EDX analysis showed average presences of Fe in this fraction Figure 3(b). The EDX spectrum Figure 3(b) also shows high Fe content which is 12.30% weight and 3.33% the atomic percentage of the Fe.

IOP Conf. Series: Materials Science and Engineering 1231 (2022) 012013 doi:10.1088/1757-899X/1231/1/012013

(a) (b)
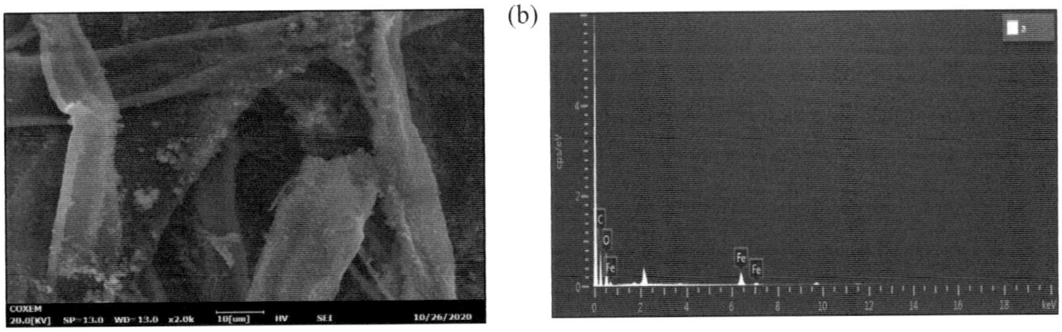

Figure 3. The SEM surface morphology micrograph of sugarcane bagasse fibre (a) surface morphology (b) The EDX pattern of bonding between the magnetic particle in the lumen of the fibre for soda process technique.

Among SEM images of irradiated magnetics sugarcane bagasse paper, the structural occur molecule agglomeration. During devolatilization, volatile matter is trapped on the surface of the particles and allows the surface to bubble initially, altering the shape of the particle and gradually bubbling out of the surface and altering the surface of the particles. Due to the low melting compound, breaking chemical bonds and melting certain compounds can cause fibre structures to separate from the original material and cause the particle to shrink in size and lead to fragmentation followed by particle agglomeration.

The EDX spectrum Figure 4 (b) shows no presences of Fe content in weight and the atomic percentage. This is due to no element of ferrite magnet that is associated with this multipurpose paper. Besides that, the presences of aluminium (Al) due to non-organic materials in it. Paper also contains nonorganic materials to improve its properties. For instance, chalk ($CaCO_3$) and kaolin clay ($Al_2Si_2O_5(OH)$). Also, titanium oxide (TiO_2) is also commonly used in paper for bleaching it. This multipurpose paper may be use bleaching during the process of papermaking .

(a) (b)
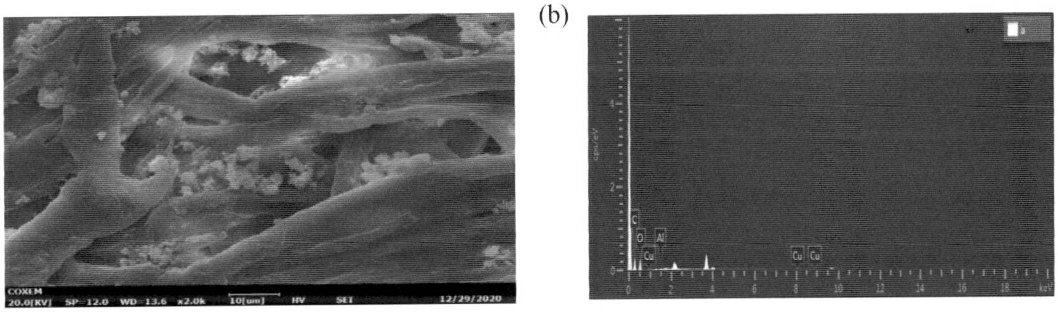

Figure 4. The SEM surface morphology micrograph of sugarcane bagasse fibre (a) surface morphology (b) The EDX pattern of bonding between the particle presents in the multipurpose paper.

Knowledge about fibre width or length important for comparing various kinds of wood and natural fibres. In cellulose based fibres, a high aspect ratio (width or length) is very important as it gives an indicator of potential strength properties. So, differences in fibre morphology may be attributed to variations in physical properties. Major differences in structure such as density, cell wall thickness, length and diameter do result in differences in physical properties. Large structural variations, such as thickness of the cell wall, length, diameter, and density may account to a result in differences in physical properties.

Elementary sugarcane bagasse fibre wall possesses a unique multilayer configuration called polylamellate structure, where every layer is reinforced with cellulose microfibrils at different angles. The microscopic structures of the sugarcane bagasse fibres are cellulose, hemicellulose, lignin and pectin. This structure determines the mechanical properties of the technical fibers and commit to the modulus and strength of the sugarcane bagasse culm.

To summarize, most of sugarcane bagasse fibre samples presented in figures above indicated that the fibers are agglomerated. It was separated from the polymer matrix and appeared to be free of any matrix interactions as well as very visible cracks and voids. Weak interfacial stability between hydrophobic polyolefins and hydrophilic cellulosic fibres was the cause from the absence of any physical interaction between the fibres and the polymer matrix.

On the other hand, the usual sugarcane bagasse fibre showed and exhibited strong interfacial adhesion as the fibers adhere better to the surfaces of the polymer matrix. This is may be due to radiation technique that cause the changes of surface morphology in the magnetics sugarcane bagasse paper.

3.2 Characterization with Fourier-transform infrared spectroscopy (FTIR)

Figure 5(a) illustrates the vibration peaks at 3335.88, 1315.88, 1029.45, 558.87 and 435.74 cm^{-1}. The presence of these vibrations peaks indicates the formation of matrix generating in the lignin that is associate with cellulose and hemicellulose. This spectrum also assigned to alcohol group respectively.

Next, figure 5(b) shows the spectrum of multipurpose paper. The spectrum of this sample shows absorption bands at 3334.69, 1316.01, 1160.19, 1027.74, 558.71 and 435.39 cm^{-1}. While the peak at 3334.69 cm^{-1}, 1160.19 cm^{-1}, and 1027.74 cm^{-1} correspond to O–H linked shearing (polysaccharides), C–O–C asymmetrical stretching (cellulose) and C-O stretching respectively. Figure 4.8(b) which is a modern magnetics sugarcane bagasse paper. The spectrum at 1.0 kGy dose spectrum exhibits the vibration peaks at 3332.64 and 2893.31 cm^{-1} correspond to O–H linked shearing (polysaccharides) and C–H symmetrical stretching (polysaccharides) respectively. The results of the characterization in Figure 6 show the entire sample exhibit the absorption range of 3400 to 2400 cm^{-1} region which shows the characteristics for stretching vibration of O-H and C-H bonds.

To summarize, the spectra of samples in Figure 5 and 6 represent absorption bands of magnetic particles and fibers. This indicates an effective interaction between ferrite nanoparticles and the cellulose network of fibers. The magnetic properties may exhibit ferromagnetic behaviour as shown by magnetization [5].

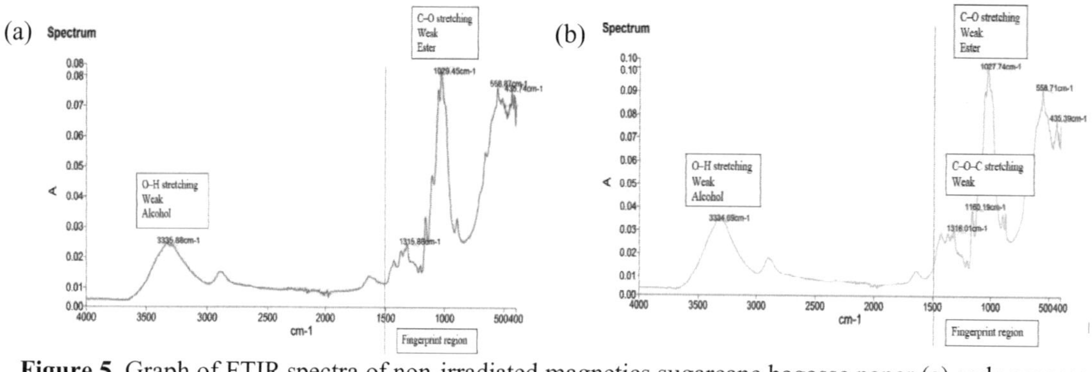

Figure 5. Graph of FTIR spectra of non-irradiated magnetics sugarcane bagasse paper (a) soda process (b) multipurpose paper.

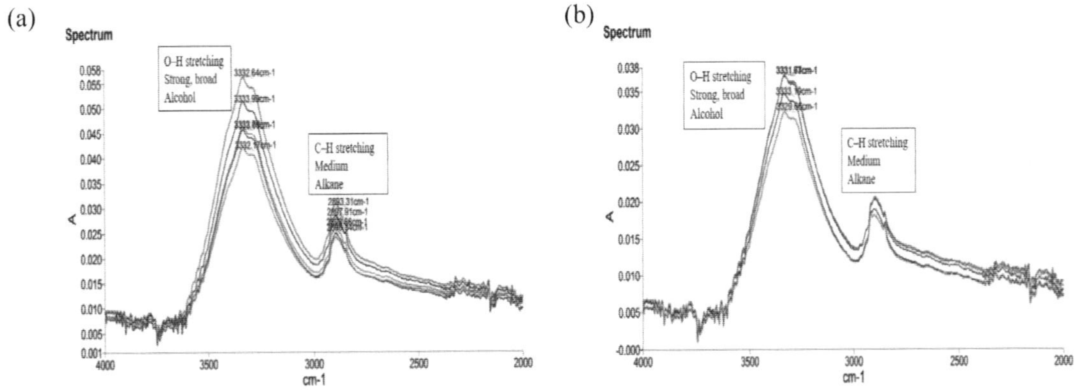

Figure 6. Graph of FTIR spectra of irradiated magnetics sugarcane bagasse paper with different doses of radiation which are 0.5kGy,1.0kGy, 1.5kGy, 2.0kGy, and 2.5kGy (a) soda process (b) multipurpose paper.

Figure 7 shown the water absorbency time for the different types of papers produced from magnetics sugarcane bagasse paper and multipurpose paper. The absorbency of magnetics sugarcane bagasse paper is quite lower than that of multipurpose paper. Its means magnetics sugarcane bagasse paper fibre can be used for packaging, printing and wrapping paper etc. By compared magnetics sugarcane bagasse paper and multipurpose paper, magnetics sugarcane bagasse paper lower absorbency than multipurpose paper [6].

The absorbency will influence the mechanical and surface properties of paper created which shows less dimensional dependability contrary to what would be expected. The quality paper needs generally excellent dimensional stability contrary to what would be expected because the structure and the quality of the paper rely upon it. Cellulose fibre can expand from 15 to 20 % from dry condition to immersion where it can cause the change is solidness. Such a change in measurement will make the dimensional stability decline cause undesirable cockling and twisting in the dimensional strength of the paper.

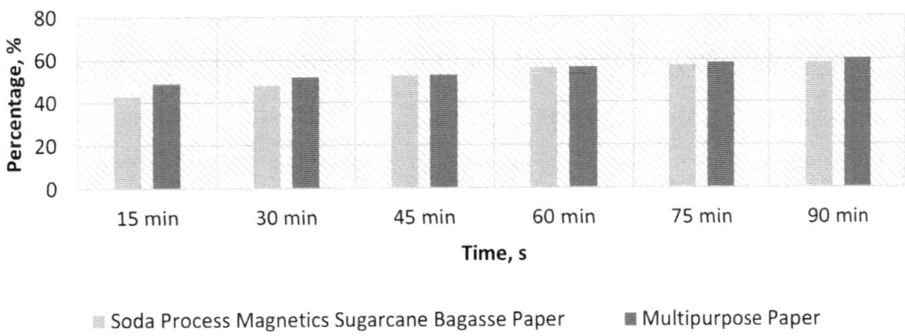

Figure 7. Graph of percentage of water absorbency between magnetics sugarcane bagasse paper and multipurpose paper.

Figure 8 shows the pH reading between non-irradiated magnetics sugarcane bagasse and multipurpose paper. If we look at the trend of pH over time, we can see that uptake soda process technique by using machine and technology are more alkaline than others. According to [7], at this point, when the pH rises as high as pH 7~10. This may due to the loss of the impurities that exist on the surface of the fibre. This can be clarified that the properties of magnetics sugarcane bagasse paper were similar with commercial paper.

The pH reading which is alkaline was due to the properties of the pulp itself and makes this magnetic sugarcane bagasse paper are suitable as a substitution as paper.

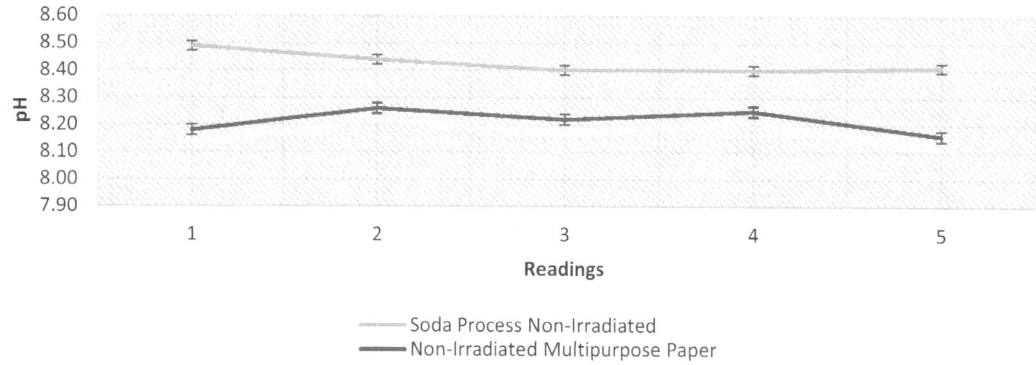

Figure 8. Graph of the pH reading between non-irradiated magnetics sugarcane bagasse paper technique with multipurpose paper.

Figure 9 illustrates the results of post-irradiation magnetics sugarcane bagasse paper with multipurpose paper. a graphical graph to interprets more about the pH reading. By comparing the pH value between irradiated magnetic sugarcane bagasse paper with multipurpose paper, their pH value is more than multipurpose paper. This can be clarified that magnetics sugarcane bagasse paper are alkaline and similar to multipurpose paper. Also, magnetics sugarcane bagasse paper can be potentially considered suitable as a substitution for paper.

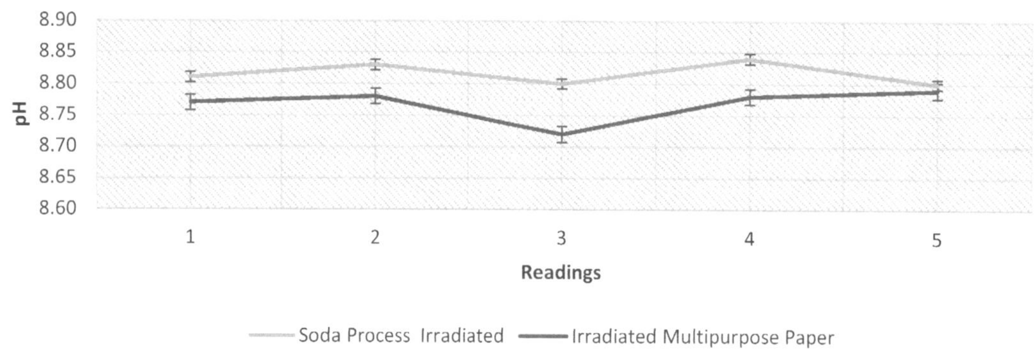

Figure 9. Graph of the pH reading between irradiated magnetics sugarcane bagasse paper and multipurpose paper

4. Conclusion

Without a doubt, the introduction of filler in this study alter the properties such as texture, strength, and opacity. Thus, ferrite (Fe) magnet as a filler enhance the paper physically as ferrite is a hard-magnetic material with distinct properties such as good mechanical hardness and chemical stability, thereby it is a convenient material for magnetic paper production. The physical changes of the paper may be attributed to thermal, structural or morphological of the fibre after the fibre treatment or irradiation process.

5. References

[1] Samariha A and Khakifirooz A 2011 *BioResources* **6** 3313.

[2] Zakaria S, Ong B H and Van de Ven G M 2004 *Colloids and Surfaces A: Physicochemical and Engineering Aspects* **251** 31.

[3] D'Almeida M L O, Barbosa P D S M, Boaratti M F G and Borrely S I 2009 *Radiation Physics and Chemistry* **78** 489.

[4] Kapoor K, Garg N, Diwan R K, Varshney L, and Tyagi A K 2017 *Radiation Physics and Chemistry* **141** 190.

[5] Piñeres O H, Salcedo E C, Herrera A P, Sánchez J H, and Quintana G C 2020 *Cellulose* **27** 3903.

[6] Al-Sulaimani K and Dwivedi P B 2017 *Int. J. Stud. Res. Technol. Manag.* **5** 16.

[7] Sutan N M, Mazlan S M, Taib S N L, Lee D T C, Hassan A, Sahari S K, and Sobuz H R 2018 In *E3S Web of Conferences* (EDP Sciences) **34** 02051.

Acknowledgement
The authors would like to thank the Universiti Tun Hussein Onn Malaysia for facilities provided.

Mechanical properties and thermogravimetric analysis of peroxide prevulcanized natural rubber latex induced by Co-60 γ radiation

Sofian Ibrahim[1,(a)], Khairiah Badri[2], Chantara Thevy Ratnam[1], Chai Chee Keong[1], Hasan Sham[1], Mohd Noor Wadi Mat Lazim [1], Noor Hasni M. Ali[1] and Khairul Hisyam Mohd Yusof[1]

[1]Malaysian Nuclear Agency, 43000 Kajang, Selangor, Malaysia
[2]School of Chemical Science and Food Technology, Faculty of Science and Technology, Universiti Kebangsaan Malaysia, 43600 UKM Bangi, Selangor, Malaysia

[a]sofian_ibrahim@nm.gov.my

Abstract. The radiation and peroxide vulcanizations of natural rubber latex are sharing the same problem which is low mechanical properties on their latex. This makes it unsuitable to be use in the production of premium latex products such as surgical glove. Moreover, there are some cases where the use of sensitizers in radiation vulcanization and activators in peroxide vulcanization tends to produced latex film with unpleasant smell and darken color respectively during the drying process. For this study, radiation of latex formulations based on 0.1 pphr of tert-butyl hydroperoxide and 0.06 pphr of potassium laurate at various radiation doses showed increment of tensile strength with increasing of radiation doses; radiation at 25 kGy produced rubber film with tensile strength of 20.7 MPa which is almost 7 times higher than control.

1. Introduction

One of the major contributors to Malaysia's national income is rubber and latex-based products. Based on Natural Rubber Statistics 2020 report [1], Malaysia's manufactured rubber goods sales from year 2008 to 2020 showed a very encouraging increment, especially on rubber gloves product (Figure 1). From the data, it is shown that on 2019 the sale of rubber gloves alone contributed RM 16.35 billion to national income, an increment of 6.9 % from previous year. Therefore, a continuous research and development in latex technology is essential to guarantee the expansion of this positive sales performance.

In latex dipped products production such as surgical gloves, it is known that natural rubber latex needs to be vulcanized prior to use in rubber glove production. At present, there are three major types of vulcanization methods being used in natural rubber latex industries; namely sulphur, radiation and peroxide vulcanizations. However, sulphur vulcanization has a major drawback in the regards to its by products, i.e. nitrosamines and nitrosatables. These unwanted by products are carcinogenic materials which may cause cancer and chemical allergies [2,3].

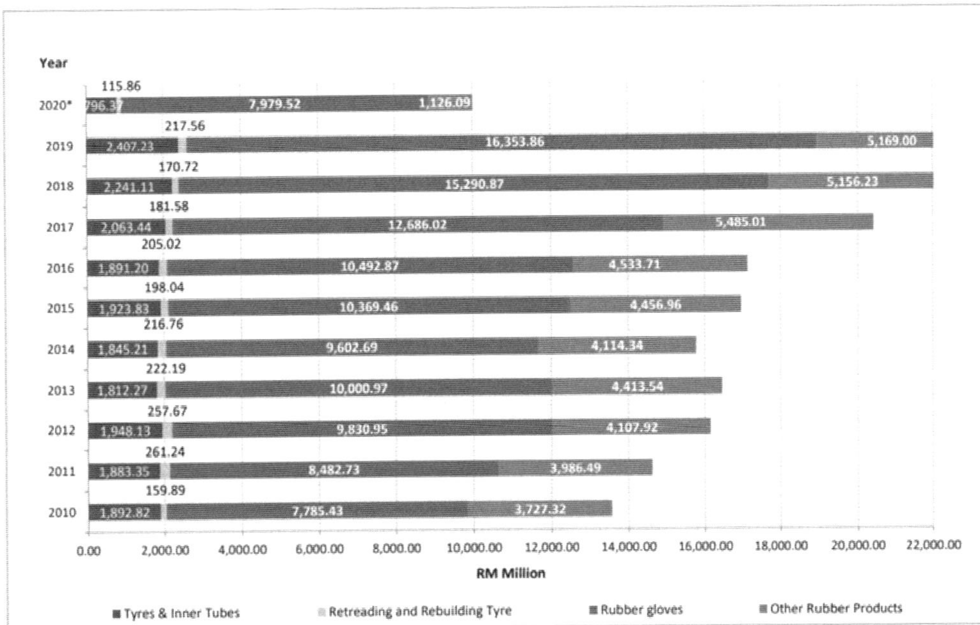

Figure 1. Sales value of Malaysia manufactured rubber goods from 2008-2020 (RM million).

The peroxide vulcanization of natural rubber latex (NRL) was introduced by the Malaysian Rubber Board (MRB) since the early 1990s. Alike the Radiation Pre-vulcanized Natural Rubber Latex (RVNRL) that was introduced by the Malaysian Nuclear Agency, peroxide vulcanization also faced the same problem which is produced low mechanical properties (tensile strength) latex. This makes it unsuitable to be use in the production of premium latex products such as surgical glove and etc. Moreover, there are some cases where the use of radiation sensitizer in radiation vulcanization and activators in peroxide vulcanization tends to produced NRL film with unpleasant smell and darken color during the drying process [4,5].

In conventional peroxide vulcanization, organic peroxide will be mixed with NRL and activators such as tetra-ethylene pentamine, fructose and hydroxyacetone before heating at 70 °C to produced prevulcanized NRL with tensile strength of approximately 18 to 22 MPa [6]. The function of these activators is to yield free radicals from peroxide to promote the formation of rubber chains crosslinking, whereas without the presence of the activator the crosslink reaction of the NRL will occur at very low rate [7].

Based on our previous study, the activator and co-agent in peroxide vulcanization are no longer needed when using gamma radiation because the high radiation energy from ionized radiation capable to promote the peroxide decomposition in NRL for the formation of rubber chains crosslinking [8]. For this study, the optimum radiation doses for vulcanization of NRL via peroxide vulcanization system are determined based on the mechanical properties (tensile strength) of the films. This study also extended to examine the effects of radiation on crosslinking density and thermal stability of the Peroxide Prevulcanized Natural Rubber Latex (PVNRL) films.

2. Materials and methods

2.1. Materials

The latex utilized in this work is a high ammonia type of latex (HA latex) supplied by Revertex (M) Pt. Ltd., Malaysia. The peroxide used were tert-butyl hydroperoxide (*t*-BHPO) supplied by Fluka, Switzerland and the stabilizer used was potassium laurate supplied by Tiarco Chemical (M) Pt. Ltd.,

Malaysia and the antioxidant used was Aquanox Lp supplied by Aquaspersion (M) Sdn. Bhd., Malaysia. These materials were used as received.

2.2. Preparation of PVNRL compounding formulations
Six kilogram (kg) latex compounding formulation for peroxide vulcanization preparation using *t*-BHPO is given in Table 1. The peroxide, stabilizer and water were first prepared into an emulsion before slowly added into the latex with gentle stirring. Once the addition of the emulsified materials was completed, the latex mixture was left stirring for three hours [8, 9]. It was then transferred into 6 separated one litre screw capped plastic container and irradiated with gamma rays from a cobalt-60 source at MINTec-Sinagama Plant, Malaysian Nuclear Agency (348778 Curie; dose rate 1.61 kGy/hr) at varying doses of 0, 5, 10, 15, 20 and 25 kGy respectively. After radiation, the latex was made into film by coagulant dipping method to measure its mechanical properties.

Table 1. Compounding formulation for *t*-BHPO vulcanization.

Materials	Part perhundred rubber (pphr)
NRL (62% Total solid content; TSC)	100
t-BHPO	0.10
Stabilizer	0.06
Antioxidant	2.50
Water	Add to 52% TSC

2.3. Measurement of tensile properties
Specimens for tensile testing were prepared using the coagulant dipping method [3, 10]. A glass plate was immersed in the coagulant and then placed in an oven at 100°C to partially dry the coagulant. It was then immersed in the latex compound for 20 s. The wet gel was allowed to consolidate at 100°C for 1 minute, and followed by leaching in distilled water at 60°C for 5 minutes. The latex film was finally dried at 100°C for 30 minutes and subjected to tensile test using Universal Testing Machine Shimadzu M703911, 50N in accordance to ASTM D412 [11]. The latex films samples were cut into dumbbell shape test pieces (Figure 2). Five samples were used for tensile test and a median value was taken as the final result.

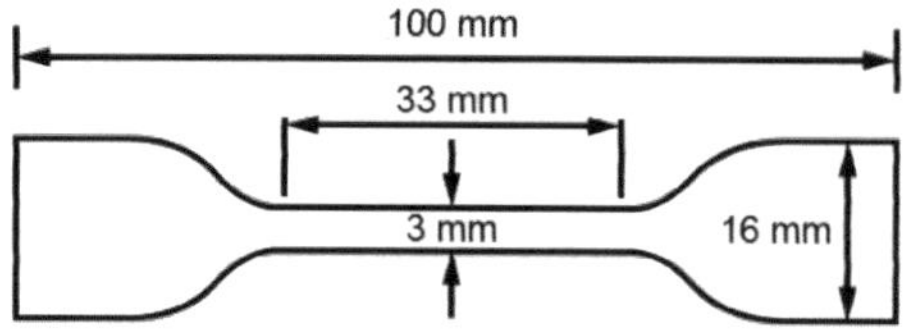

Figure 2. Dimension of dumbbell cut

2.4. Determination of gel content
The gel content of the crosslinked samples was determined by the extraction of samples in toluene for 8 hours using Soxhlet apparatus as required by [12, 13]. The extraction samples were dried in an oven at 70°C till constant weight was achieved. The gel fraction was calculated as Eq. (1):

IOP Conf. Series: Materials Science and Engineering 1231 (2022) 012014 doi:10.1088/1757-899X/1231/1/012014

$$\text{Gel content, \%} = \frac{w_1}{w_0} \times 100 \quad (1)$$

where w_0 and w_1 are the weights of the dried samples before and after extraction, respectively.

2.5. Thermogravimetric analysis (TGA) and Differential thermal analysis (DTA)

TGA and DTA analysis was performed using the Perkin Elmer model of Pyris 1 Thermal Analyzer where heating process was performed in a nitrogen environment at 30-650 °C temperature range with 10 °C / min heating rate.

3. Results and discussion

3.1. Effect of radiation doses on tensile strength of RVNRL

The suggested mechanisms for decomposition of peroxide by radiation were shown in Figure 3. In such a mechanism, RH, B and RR will represent natural rubber latex, peroxide and prevulcanized natural rubber latex.

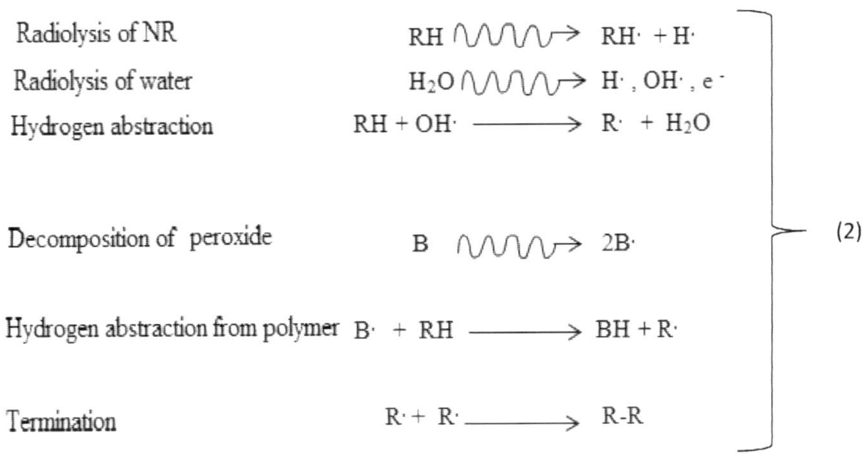

Figure 3. Mechanism for decomposition of peroxide by radiation

Parameters such as tensile strength and modulus values are very important in the production of latex dipped products. This is because some products have to meet certain standards. For example, the minimum tensile strength for surgical glove is 24 Mega Pascal (MPa) as required by the ASTM D3577-01a standard [14], while the minimum tensile strength for examination glove is 18 MPa as required by the ASTM D3578 – 05 [15]. However, from our previous study mixture of NRL and 0.1 pphr *t*-BHPO irradiated at 12 kGy only capable to produced PVNRL with 15 MPa tensile strength [8]. Thus, investigation on the precise amount of radiation dose is very important because it can give major impact on the mechanical properties of the latex [2].

Since the films are cast after radiation process, the particles coalesce together into transparency film during evaporation process of emulsion medium; the interaction between particles determines tensile strength and elongation at break of the film. Table 2 gives the mechanical properties of PVNRL samples that have been prepared by gamma radiation at various doses and tested as required by ASTM D412 standard.

Table 2. Mechanical properties of PVNRL irradiated at various dose (median value from tensile test).

Sample	Radiation dose (kGy)	Modulus @ 500% (MPa)	Modulus @ 700% (MPa)	Tensile strength (MPa)
PVNRL (Control)	0	0.31	1.52	3.1
PVNRL	5	0.83	2.85	12.4
PVNRL	10	0.97	3.19	17.5
PVNRL	15	1.51	4.54	19.0
PVNRL	20	1.35	4.28	20.0
PVNRL	25	2.23	5.06	21.7

It was observed that the tensile strength, modulus at 500 % and modulus @ 700 % of the films increase continuously with increasing radiation dose. The rubber film obtained from radiation at 25 kGy had a tensile strength, modulus at 500 % and modulus @ 700 % of 21.7, 2.23 and 5.06 MPa respectively; which is more than six times increment compared to control.

The increasing of the tensile strength of PVNRL is suggested due to the increasing of the intraparticle crosslink density (chemical crosslinking) caused from increasing radiation dose which leads to the formation of a three-dimensional network or rubber matrix [16]. Besides, the interparticle entanglement (physical crosslinking) also suggested to increase with the increasing dose because of the interparticle entanglements are depending on the free rubber chain ends at the surface of each latex particle, whereby these chains interpenetrate during film formation and contribute to the strength of the film by means of entanglements. [2].

3.2. Effect of radiation doses on gel content of PVNRL

The gel content or percentage of crosslinking is determined based on ASTM D3616-95 [12] requirement. The effect of radiation doses on gel content of PVNRL are shown in Table 3. From the table, it was observed that the extent of gel formation increases with the increasing of radiation dose, indicating increases in crosslink density of the polymer. At 25 kGy of radiation dose, PVNRL film produced 84.5 % crosslink percentage compared to the control sample at; 35.9 %.

Table 3. Effect of radiation doses on gel content of PVNRL.

Sample	Radiation dose (kGy)	Gel content (%)
PVNRL (Control)	0	35.9
PVNRL	5	69.1
PVNRL	10	75.3
PVNRL	15	82.6
PVNRL	20	83.9
PVNRL	25	84.5

3.3. TGA analysis

TGA analytical technique was used to determine a material's thermal stability for PVNRL that was prepared at various radiation doses (Figure 4). When NRL are crosslinked by peroxides and radiation, C-C bonds are formed between individual rubber chains. The C-C bond is stronger and more thermal stable than the C-S bond formed by sulphur vulcanization.

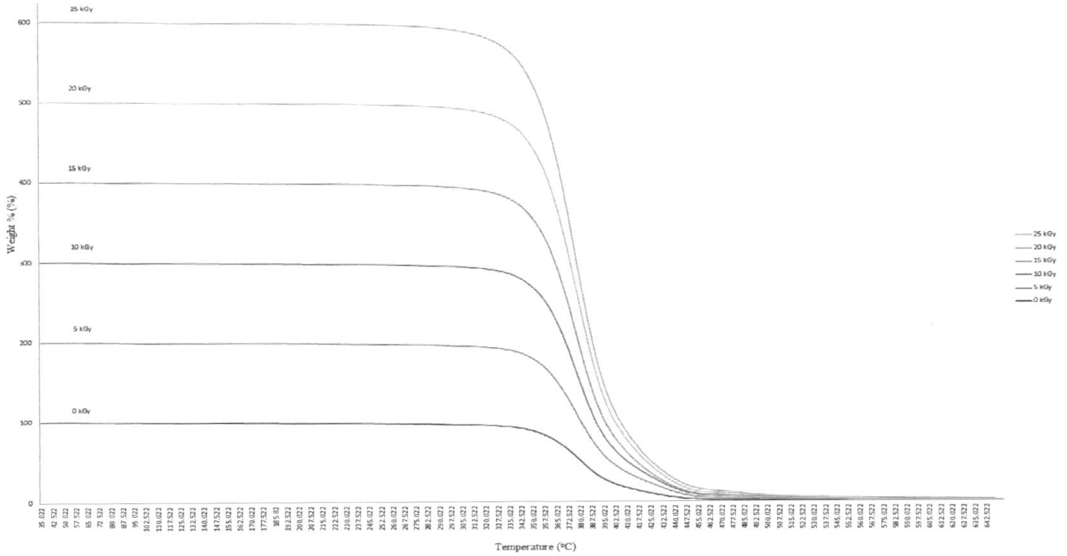

Figure 4. TGA thermograms for PVNRL prepared at various radiation doses.

From the TGA analysis that was conducted on all PVNRL samples, it was observed that the weight loss of the PVNRL samples occurred in the temperature range of 300–540 °C corresponding to a degradation of the non-volatile residues (Table 4). Besides, T_{deg} for the samples are around 378 °C. From the weight loss curves, a clearer picture of the effect of radiation doses on the thermal stability of the latex has been observed. The thermograms clearly show a less steep gradient for the entire PVNRL sample that was prepared from 5 to 25 kGy compared to the control (0 kGy). These results were attributed to an enhancement of the C-C crosslink density of PVNRL as consequence of the dose effect; whereas increasing of the C-C bond density also increase the thermal stability of the latex [17].

Table 4. Decompositions temperature of PVNRL at various radiation doses

Radiation dose (kGy)	T_0 (°C)	T_{deg} (°C)	T_{end} (°C)
0	350	378	475
5	353	379	505
10	353	378	510
15	352	378	520
20	353	378	528
25	353	378	535

4. Conclusions

Peroxide prevulcanized natural rubber latex (PVNRL) with promising high tensile strength can be prepared by induced with Co-60 γ radiation. Radiation of latex formulations based on 0.1 pphr of tert-butyl hydroperoxide (*t*-BHPO) and 0.06 pphr of potassium laurate as stabilizer at various radiation doses showed the increment of the tensile strength as the increasing of radiation doses. At radiation dose of 25 kGy, rubber film with tensile strength of 20.7 MPa and crosslink percentage of 84.5 % has been successfully produced.

5. References

[1] Malaysian Rubber Board 2020 Natural rubber statistics 2020 (January- March) *Credible rubber pricing systems* viewed 15 October 2020. < http://www.lgm.gov.my/nrstat/Statistics%20Website%202020%20(Jan-Mar).pdf>.

[2] Makuuchi K 2003 *An introduction to radiation vulcanization of natural rubber latex* (Bangkok:T.R.I. Global Co. Ltd).

[3] Sofian I, Badri K, Chantara T R, Noor Hasni M A 2018 Enhancing mechanical properties of prevulcanized natural rubber latex via hybrid radiation and peroxidation vulcanizations at various radiation doses *Radiat. Effects and Defects in Sol.* **173**(5-6) p427-434.

[4] Wan Manshol W Z, Mohid N and Razali M Y 1995 RVNRL A potential material in latex dipped products manufacturing *Radiat. Phys. Chem.* **46**(4-6) p1019-1023.

[5] Roslim R, Jefri J. Manroshan Singh J S, Siti Noor Suzila M U H, Amir Hashim M Y 2015 *Characterizing physical properties of peroxide vulcanized natural rubber latex films* (Switzerland: Trans Tech Publications Ltd.) **1134** p236-242.

[6] Ma'zam M S, Pendle T D, Blackley D C 1990 peroxide prevulcanization of natural rubber late. *J. Nat. Rubber Res.* **5** (1) p27-40.

[7] van der Hoff B M E 1963 Reactions between peroxide and polydiolefins *Ind. Eng. Chem. Prod. Res. Dev.* **2** (273).

[8] Sofian I, Badri K, Chantara T R, Noor Hasni M A, Chai C K, Mohd Noorwadi M L 2018a Preliminary study on Peroxide vulcanization with gamma radiation *J. Sains Nuklear Malaysia* **30**(2) p39–46.

[9] Wan Manshol W Z and Chai C K 2000 Preparation and mechanical properties of low proteins radiation vulcanised natural rubber latex *MINT R&D Seminar 2000* (Bangi, Selangor, Malaysia: MINT).

[10] Pairu I, Rusli D and Wan Manshol W Z 2016 Thermal and mechanical properties of gamma-irradiated prevulcanized natural rubber latex/low nitrosamines latex blends *Radiat. Effects and Defects in Sol.* **171**(1-12).

[11] ASTM D412 2016 *Standard test methods for vulcanized rubber and thermoplastic elastomers—tension* (West Conshohocken, PA: ASTM International).

[12] ASTM D3616-95 2014 *Standard test method for rubber—determination of gel, swellingindex, and dilute solution viscosity* (West Conshohocken, PA: ASTM International).

[13] Jayasuria M M, Makuuchi K, Yoshi F 2001 Radiation vulcanization of natural rubber latex using TMPTA and PEA. *Euro. Polym. J.* **37** p93-98.

[14] ASTM D3577-01a 2001 *Standard specification for rubber surgical gloves* (West Conshohocken, PA: ASTM International).

[15] ASTM D3578 – 05 2015 *Standard specification for rubber examination gloves* (West Conshohocken, PA: ASTM International).

[16] Kruželák J, Sýkora R, Hudec I 2017 Vulcanization of rubber compounds with peroxide curing system *Rubber Chem. and Technol.* **90** (1) p60–88

[17] Kim D Y; Park J W; Lee D Y, Seo K H 2020 Correlation between the crosslink characteristics and mechanical properties of natural rubber compound via accelerators and reinforcement. *Polym.* **12**(9).

Acknowledgements

The authors would like to thank RAYMINTEX Plant, Malaysian Nuclear Agency for financial and technical supports of this work and the Revertex (M) Sdn. Bhd., Malaysia for supplying HA latex to be used in this research.

IOP Conf. Series: Materials Science and Engineering 1231 (2022) 012015 doi:10.1088/1757-899X/1231/1/012015

Time dependent reliability analysis for a critical reactor safety system based on fault tree approach

Ahmad Hassan[1,a)], Zahira Mokhtar[1], Mazleha Maskin[2]

[1]Faculty of Applied Sciences and Technology, Universiti Tun Hussein Onn Malaysia, 86400 Muar, Johor, Malaysia.
[2]Malaysian Nuclear Agency (Nuclear Malaysia), Bangi, 43000 Kajang, Selangor, Malaysia

[a]hassan@uthm.edu.my

Abstract. Loss of coolant in the operation of any nuclear power plant will eventually become the primary source of hazard in the sequence of events leading to reactor core uncovery. Subsequent failure in removing the nuclear decay heat and preventing the core uncovery will further lead to loss of coolant accident (LOCA). In conjunction to this safety concern, it is crucial that the reliability of the plant emergency core cooling system is systematically and critically analysed. This article presents a case study on the time dependent reliability analysis for a safety injection system (SIS) of an advanced pressurized water reactor, based on the failure mode and effect analysis (FMEA) and fault tree (FT) analysis approach. The identified generic data for component reliability are carefully reviewed and used in this study. Based on the base case model, sensitivity and importance measure analysis for basic events are performed and the outcomes gained are presented and discussed. From the analysis, it is shown that the safety injection pumps of the SIS contribute significantly to the reliability of the system. In the short (at 0.5 hour) and long (7.0 hours and 72.0 hours) run, safety injection pumps are critical and influence the reliability of the SIS the most. The SIS's FT logic model that has been developed and calculated shows the usability of the FMEA and FT approach that are implemented in analysing SIS time dependent reliability.

1. Introduction

The history of nuclear power has expanded almost nearly a century and within this timeline, the world has witnessed the raise and decline of its demand. Among many other factors to this declination is the safety issues, that were usually seen argued and discussed intensively after the occurrence of major accident at a nuclear power plant. On March 2011, Fukushima Daiichi Nuclear Power Plant marked another history of nuclear disastrous event.

The safety of a nuclear power plant is analyzed as early as in the stage of reactor design, and after that are the design licensing, site selection, construction, commissioning and up to the decommissioning which involve and follow strict and stringent procedures. However, Fukushima event was so far an exception since the actual sequence of event that has propagated was complex, mainly due to the occurrences of multiple simultaneous initiating events. Since then, the nuclear safety community has initiated the efforts on considering such complex occurrence and it is to be carefully analyzed, especially by means of the probabilistic approach[1].

For the year of 2020, it is reported that the share of nuclear energy to the global primary energy generation is around 4.3%[2]. For any particular nuclear power plant, it is worth to realize that the energy is originated from the fission in the nuclear fuel core. However, during any plant incidents that involve the loss of integrity at the primary coolant boundary, the core will eventually become the source of hazard in the anticipated sequence of events that lead to core uncovery. Failure in the recovery of primary coolant and preventing the core uncovery will lead to the sudden increase in the fuel cladding temperature, and thus damaging the fuel rods and fuel assemblies[3]. Thus, in every plant, a robust and reliable emergency cooling function must be critically designed and analyzed.

This article is about a case study on a time dependent reliability analysis of a safety injection system (SIS) of an advanced pressurized power reactor during post-large break loss of coolant accident (LOCA), performed based on fault tree (FT) analysis approach. Technically, this involves the familiarization of the system and definition of the accident scenario, failure mode and effect analysis (FMEA), FT modeling and FT quantification.

The next section of this article briefs the design feature of the Advanced Power Reactor 1400 (APR1400), and later the structure of SIS is explained in detail. The summary of large break LOCA accident phenomena that were analyze by others and the scope of analysis in the current study are addressed. In Section 3, the failure mode and effect analysis (FMEA) is explained. Mathematical fundamentals for the fault tree analysis are explained and the use of a computer software is addressed. For component reliability data, generic data available from literature are used. The result of this study is presented in Section 4. Based on the base case model, sensitivity and importance measure analysis are performed and the insights gained are further discussed. Finally, Section 5 summarizes the outputs and insights of the current study and recommendation for future studies are addressed as well.

2. Overview of APR1400 reactor and the safety injection system

The reactor that is considered in this study is the APR1400 reactor, which design evolved from the historic Korean Optimized Power Reactor 1000 (OPR1000) reactor[4,5]. The plant is an advanced plant, which was eventually granted the standard design certification by United States Nuclear Regulatory Commission (USNRC) in 2018[6], whereas long before that in 2001, by the Korean regulation authority[7]. As of in the middle of 2020, there are 2 APR1400 plants that have been connected to grid in Korea while 8 other plants are under construction in Korea and United Arab Emirates (UAE)[8].

The electrical power output of the reactor is 1450MWe while the thermal power output is 4000 MW. This thermal power which is generated inside the reactor pressure vessel (RPV) is transferred out via two coolant loops system. Each loop consists of one hot leg and two cold legs, one steam generator and two recirculation pumps. A pressurizer is connected to one of the hot legs. Major engineered safety features of the plant to cope with the severe accident are the auxiliary feedwater system (AFWS), containment spray system(CSS) and SIS[7]. As depicted in **Figure *1***, a single SIS train basically consist of an In-containment Refueling Water Storage Tank (IRWST), motor operated valves (MOV), check-valves (CV) and a Safety Injection Tank (SIT) [9]. For the APR1400, there are 4 SIS trains in total; 2 trains are connected to direct vessel injection (DVI) nozzles whereas the other 2 trains are designed with connection to both the DVI nozzles and the hot legs(HL), as shown in **Figure *1***[8]. The location of these DVI nozzles is above their respective cold leg connections on the RPV. It was reported that the small and medium break LOCA events contribute about 20.1% to the core damage frequency[10]. After the initiation of large break LOCA events, borated water from the IRWST is injected into the RPV with the safety injection pump (each with 50% capacity), and at the same time rapidly from the safety injection tank[11]. Meanwhile, for the long term cooling after a large break LOCA, borated water must be continuously injected into the RPV via the latter type trains i.e. Train 1A and Train 2B. To achieve the 100% pump capacity during this long term cooling, both of the trains must be available[10]. The reliability of this cooling function by the SIS is modelled and analysed in the next sections.

3. Methodology

FMEA involves the identification and screening of the relevant failure mode of the SIS's subsystems and components and their effects on the defined functionality of the SIS during the post-large break LOCA long term cooling. These outcomes of FMEA are then arranged deductively as basic events or intermediate events, leading to top event of the intended FT model.

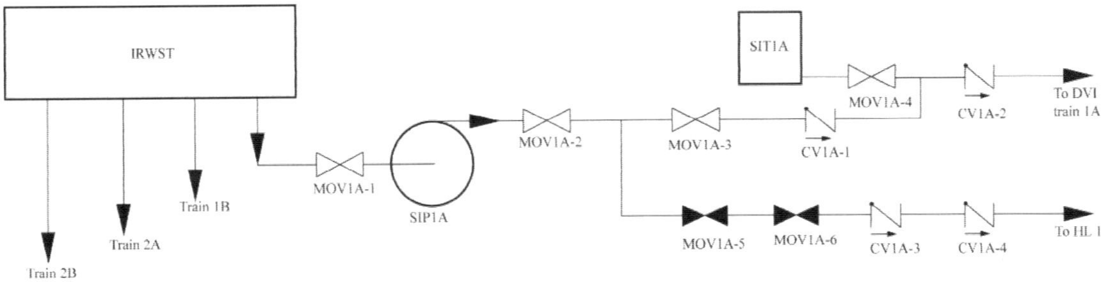

Figure 1. Schematic diagram of the SIS (Train 1A) [8]

Unavailability and unreliability share the same technical definition, except that the former usually used to refer to repairable type component[12]. Assuming that the time to failure, t for most of the physical component follows the exponential distribution (constant failure rate), the time dependent unreliability, $Q(t)$ of a component is given by[13]:

$$Q(t) = 1 - e^{-\lambda t} \tag{1}$$

λ is the component failure rate which taken from available literature. To calculate the FT top event probability, the probability of components in a cut set are multiplied together, i.e.:

$$C_i(t) = \prod_{k=1}^{n_{ik}} Q_k(t) \tag{2}$$

$C_i(t)$ is the probability of the ith cut set, n_{ik} is the number of component in the ith cut set and $Q_k(t)$ is the probability (unreliability) of the kth components in the ith cut set at specific time t. The FT top event probability Q_{top} is gained by adding together probability of the all n cut sets[14]:

$$Q_{top}(t) = \sum_{i=1}^{n} C_i(t) \tag{3}$$

Thus, $Q_{top}(t)$ represent the unreliability of the SIS for post large break LOCA long term cooling

In this study, the graphical construction and calculation of the FT is performed by using RISK SPECTRUM computer software[15]. The software generates cut sets for the constructed FT based on the gates used, simplifies the cut sets according to the Boolean rules for the generation of minimal cut sets (MCSs). The probability of each MCSs and the FT top event is calculated according to equation (2) and (3). Based on the base case model output, the Fussell-Vesely (FV) importance is performed to evaluate the fractional contribution of the basic events to the SIS failure[16] and the sensitivity of a generic operator recovery action the top event probability is analyzed.

4. Results and discussion

4.1. FMEA

The outcomes of FMEA for the active components in Train 1A are summarized in **Table 1** below. Although not included in the table, the outcome of FMEA for Train 2B is similar to that of for Train 1A, due to the trains' similarity. The success criteria for each train is: cooling water is injected into the PRV via DVI nozzle and the hot leg. These components' reliability parameter (mean value) that was selected from NUREG/CR-6928[17], based on the description and failure mode are also included in the table. The unreliability of component with probability upon demand parameter is treated as $Q(t)$=probability (constant).

IOP Conf. Series: Materials Science and Engineering 1231 (2022) 012015 doi:10.1088/1757-899X/1231/1/012015

Table 1 FMEA for Train 1A.

No	Component	Failure mode	Effects/ Consequences	Mean value (Probability/ Failure rate [h^{-1}])
1	Safety injection pump SIP1A	Fail to start (FTS) and fail to run (FTR). Status during normal condition: standby	No flow/ Loss of flow from IRWST	FTS:1.47E-3 FTR:3.78E-4 + 5.80E-6 [h^{-1}]
2	Motorized valve MOV1A-1	Status during normal condition: open	(inadvertent/spurious closure is not modelled)	-
3	Motorized valve MOV1A-2	Status during normal condition: open	(inadvertent/ spurious closure is not modelled)	-
4	Motorized valve MOV1A-3	Status during normal condition: open	(inadvertent/ spurious closure is not modelled)	-
5	Motorized valve MOV1A-4	Fail to close (FTC). Status during normal condition: open	Water flows into SIT1A, reducing the flow capacity that going into DVI	FTC: 1.07E-3
6	Motorized valve MOV1A-5	Fail to open (FTO). Status during normal condition: close	No flow into RPV via HL1	FTO: 1.07E-3
7	Motorized valve MOV1A-6	Fail to open (FTO). Status during normal condition: close	No flow into RPV via HL1	Same as above; FTO: 1.07E-3
8	Check valve CV1A-1	Fail to open (FTO)	No flow into RPV via DVI	FTO: 1.03E-5
9	Check valve CV1A-2	Fail to open (FTO)	No flow into RPV via DVI	Same as above; FTO: 1.03E-5
10	Check valve CV1A-3	Fail to open (FTO)	No flow into RPV via HL1	Same as above; FTO: 1.03E-5
11	Check valve CV1A-4	Fail to open (FTO)	No flow into RPV via HL1	Same as above; FTO: 1.03E-5
12	IRWST	Rupture	Currently, the possible hazard the leads to this failure is not explicitly considered, thus its failure is not modelled	-
13	SIT1A	Rupture	Same as above	-
14	Pipes	Rupture	Same as above	-

4.2. SIS long term cooling fault tree

As notified in Section 2, the long term cooling after the large break LOCA requires continuous water injection by both Train 1A and Train 2B. Thus, with the outcome of FMEA in **Table *1***, failure of this function can be represented by the main FT shown in **Figure** *2*. To keep the size of the main FT at a manageable size, the events under Train 2B are transferred to a separate FT. Although not presented here, the separated FT of the Train 2B is identical to Train 1A, identifiable in **Figure *2*** under the gate @SIS A1 & 2B-2.

Table 2 lists all the minimal cuts sets (MCSs) of events that can lead to the failure of post-large break LOCA long term cooling by the SIS at time $t = 0.5$ hour, with the top event unreliability $Q(t=0.5) = 9.78E\text{-}3$. The list is ranked by the cut set's probability. None of the MCSs consist of multiple events. Thus, any single component failure event will directly prevent the SIS in reaching the prescribed success criteria. The top in the list (1st and 2nd) are the failure of safety injection pumps to start. After a sufficiently longer time i.e. 5.0 hours, the top in the list (1st and 2nd) are the failure of safety injection pumps to run (the MCS for list for $Q(t=5.0)$ is not presented here).

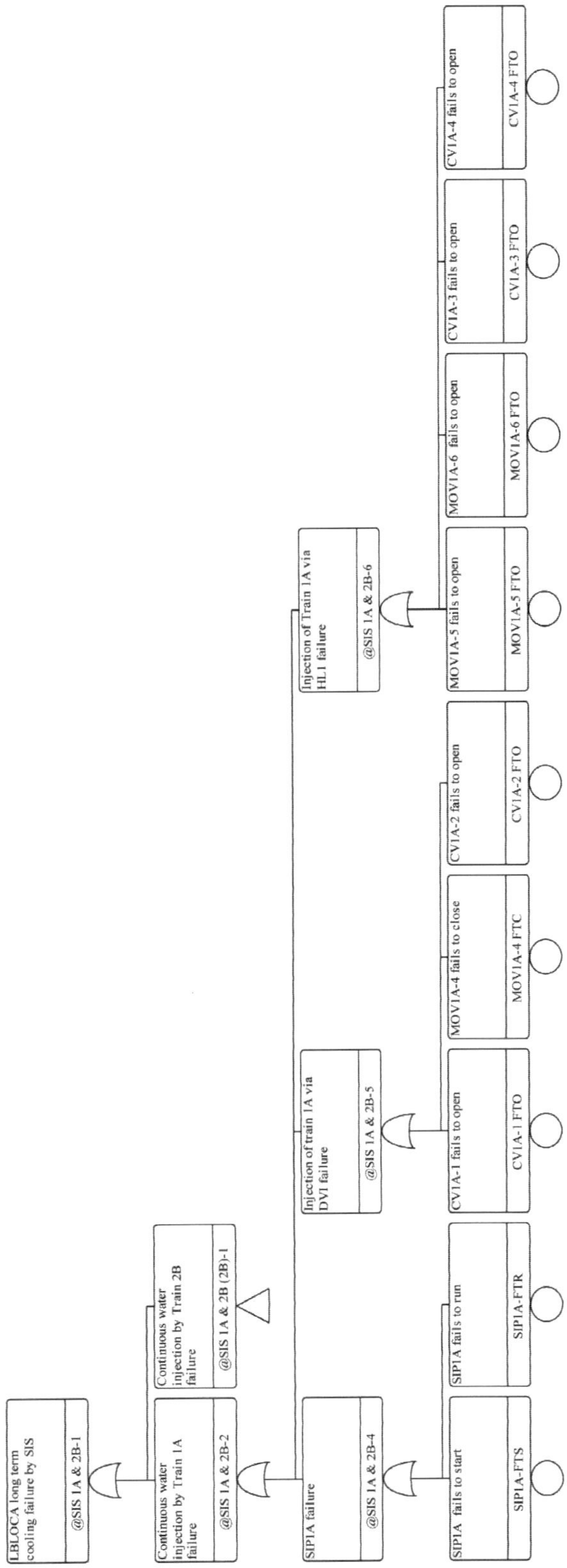

Figure 2. FT for SIS long term cooling

Table 2 MCS list with Q(0.5)= 9.78E-3

No.	Probability	%	Event 1	Event 2
1	1.47E-03	15	SIP2B-FTS	-
2	1.47E-03	15	SIP1A-FTS	-
3	1.07E-03	10.9	MOV2B-4 FTC	-
4	1.07E-03	10.9	MOV1A-5 FTO	-
5	1.07E-03	10.9	MOV1A-4 FTC	-
6	1.07E-03	10.9	MOV2B-5 FTO	-
7	1.07E-03	10.9	MOV1A-6 FTO	-
8	1.07E-03	10.9	MOV2B-6 FTO	-
9	1.92E-04	1.96	SIP2B-FTR	-
10	1.92E-04	1.96	SIP1A-FTR	-
11	1.03E-05	0.11	CV1A-3 FTO	-
12	1.03E-05	0.11	CV2B-4-FTO	-
13	1.03E-05	0.11	CV1A-4 FTO	-
14	1.03E-05	0.11	CV1A-2 FTO	-
15	1.03E-05	0.11	CV2B-1-FTO	-
16	1.03E-05	0.11	CV2B-3-FTO	-
17	1.03E-05	0.11	CV2B-2-FTO	-
18	1.03E-05	0.11	CV1A-1 FTO	-

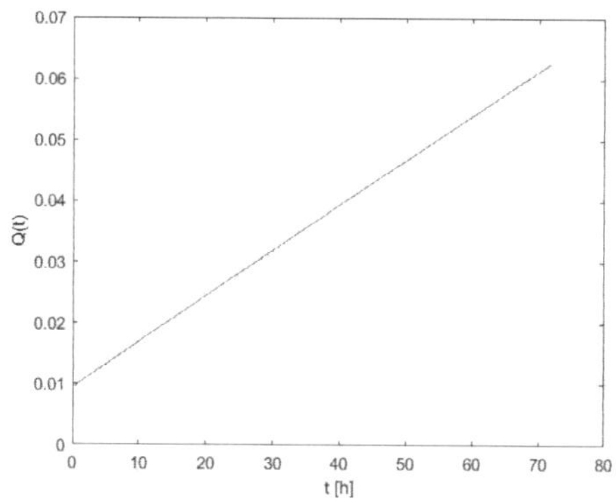

Figure 3. Unreliability of SIS with time

4.3. Time dependent reliability

Figure 3 shows the unreliability increment of the SIS as function of time. At 72nd hours, the unreliability reaches approximately 6 times of the initial unreliability (complement to this, the reliability decreases). This is observed due to the reliability of the individual SIP1A and SIP2B, which decrease by time, whereas other components which remain with constant failure probability. Thus, it can be strongly suggested that other mitigating measure to recover the long term cooling such as operator recovery actions is worth to be considered.

4.4. IM and sensitivity analysis

In this current analysis, the definition of FV importance of a basic event is the ratio of reduced unreliability of the SIS when the basic event is completely reliable (i.e. no failure at all) to the base case unreliability. The output shows that the FV importance for both SIPs fail to run is 1 and this indicates that the reduction of the SIS unreliability is 100%, if both pumps are running without any failure. For other basic events, their values are less than 1.0E-3. Thus, the pumps are the dominant contributor to the failure of SIS. If human action with nominal performance (human error probability = 0.011[18]) is considered for the recovery of the system, for the total duration of 72 hours, the mean unreliability (obtained by integration and time averaging of $Q_{top}(t)$[15]) of the system decrease from 3.63E-2 to 3.99E-4. This considerably improve the reliability of the system and therefore it is crucial that the plant operators are well trained and prepared for wide range of reactor abnormal conditions.

5. Conclusion

This study has performed a time dependent FT analysis on the APR1400's SIS for post-large break LOCA long term cooling. To this end, FMEA was inevitable for the systematic construction of the FT, and generic data originated from the similar industry have been adopted. It is shown that the reliability of the SIS degrades during its mission and mitigating measures such as human recovery actions demonstrated a considerable reliability improvement. In the short (at 0.5 hour) and long (7.0 hours and 72.0 hours) run, safety injection pumps are critical and influence the reliability of the SIS the most. Certainly, the total failure of the SIS must be followed by the next phase accident management that has been carefully understood and planned. Even though the significance of safety injection pumps was considerably highlighted, the present study has omitted other important basic events such as human errors, electrical power failures, passive components (IRWST and SIT), instrumentation and control (I&C) failures as well as the common cause failure events from the consideration. In addition, the uncertainty values of the component reliability parameter were neglected. Therefore, these topics would be the concern of our future works toward the wider scope of system analysis under the standard framework of PSA.

6. References

[1] Kwag S, Ha J G, Kim M K and Kim J H 2019 Development of effcient external multi-hazard risk quantification methodology for nuclear facilities *Energies* **12** 1–25.
[2] BP 2020 *Statistical Review of World Energy 2020 | 69th edition*
[3] Zinkle S J, Terrani K A, Gehin J C, Ott L J and Snead L L 2014 Accident tolerant fuels for LWRs: A perspective *J. Nucl. Mater.* **448** 374–9.
[4] Lee S S, Kim S H, Suh K Y 2009 The design features of the advanced power reactor 1400 *Nucl. Eng. Technol.* **41** 995–1004.
[5] Zubair M, Al Suwaidi R R and Al Souqi A A 2021 Behavior of Emergency Core Cooling System (ECCS) during the early stage of Loss of Coolant Accident (LOCA) for APR 1400 with Flownex software *Prog. Nucl. Energy* **141** 103949.
[6] Shamsi A 2018 The Peaceful Nuclear Energy Program in the United Arab Emirates : Background and history *Nucl. Law Bull.* **2018** 73–83.
[7] Kim H-G 2011 The Design Characteristics of Advanced Power Reactor 1400 *ICONE-13, Beijing, China.*
[8] KEPCO Engineering & Construction Company 2020 *Status Report – APR1400 (KEPCO E & C / KHNP).*
[9] Rezk O, Jung J and Lee Y 2016 Application of Reliability Centered Maintenance Strategy to Safety Injection System for APR1400 *J. Korea Soc. Syst. Eng.* **12** 41–58.
[10] Yun J, Kim T and Kim J 2015 Verification of SAMG entry condition for APR1400 *Ann. Nucl. Energy* **75** 404–12.
[11] Multinational Design Evaluation Programme 2018 *MDEP Technical Report TR-APR1400-01*
[12] Dutuit Y and Rauzy a. 2005 Approximate estimation of system reliability via fault trees *Reliab. Eng. Syst. Saf.* **87** 163–72.
[13] Caldarola L 1977 Unavailability and failure intensity of components *Nucl. Eng. Des.* **44** 147–62
[14] Rasmuson D M 1992 A comparison of the small and large event tree approaches used in PRAs *Reliab. Eng. Syst. Saf.* **37** 79–90.
[15] Scanpower AB 2012 RiskSpectrum Analysis Tools Theory Manual.
[16] Van der Borst M and Schoonakker H 2001 An overview of PSA importance measures *Reliab. Eng. Syst. Saf.* **72** 241–5.
[17] Eide S A, Wierman T E, Gentillon C D, Rasmuson D M and Atwood C L 2007 *Industry-Average Performance for Components and Initiating Events at U.S. Commercial Nuclear Power Plants*
[18] Whaley A M, Kelly D L, Boring R L and Galyean W J 2011 SPAR-H Step-by-Step Guidance *INL/EXT-10-18533.*

Acknowledgement
This research was supported by the Ministry of Higher Education (MOHE) through Fundamental Research Grant Scheme (FRGS/1/2019/TK05/UTHM/03/1).

IOP Conf. Series: Materials Science and Engineering 1231 (2022) 012016 doi:10.1088/1757-899X/1231/1/012016

Nuclear fuel materials and its sustainability for low carbon energy system: A review

Muhammad 'Adli Nor Azman[1], Nur Syazwani Mohd Ali[1, a)], Muhammad Syahir Sarkawi[1], Muhammad Arif Sazali[1], Nor Afifah Basri[1]

[1] Department of Energy Engineering, School of Chemical and Energy Engineering, Universiti Teknologi Malaysia, 81310 Johor Bahru, Malaysia

[a]nsyazwani@utm.my

Abstract. World energy generation for electricity is still dependent on fossil fuels since it is more reliable and secure than the current intermittent renewable energy systems. Although the integration of renewable energy as an energy mix is in progress, still it could not be able to replace fossil fuels. Dependency on fossil fuels will not only contribute to severe climate change but will also degrade future generation quality of life. Hence, the solution to quandary is by integrating nuclear power plants with those of renewable energy such as solar and wind to meet the energy demand and to ensure sustainability of energy source. The current operating nuclear power plants in the world use the concept of water-cooled reactors. It was designed so that the fast neutrons born from fission reactions were slowed down in the moderator to allow other fission reactions events in sustainable chain reactions. Besides, the slow neutrons with low energy is a favourable reactor feature for safe and efficient operation. The common types of nuclear fuel materials in water-cooled reactors are enriched uranium dioxide and natural uranium contained in nuclear fuel elements. After it has been used, the fuel elements will be stored as spent fuel. Prolonged storage of used nuclear fuels will make the volume of nuclear waste high and become hard to manage after a long period of storage. An effort to reprocess the spent fuel as to extract fissile and fertile material to be used in nuclear fuels usually was undertaken to reduce the waste volume. However, this process may lead to an undesirable proliferation of nuclear material. In this review article, research on the advancement of nuclear fuel materials will be discussed based on the reduction method of the nuclear spent fuel volume and radiotoxicity, as well as to study its sustainability for the future low carbon energy system.

1. Introduction

Climate change has pushed the world to harness a low carbon energy system. In 2019, 84.3% of the global energy consumption was produced by fossil fuels, 4.3% from nuclear and the rest of it from various renewable energy sources [1]. According to Our World in Data (2021), the trend of carbon dioxide (CO_2) emission was increasing since 1900 but reported a slow declining trend in 2019 due to the inclusion of 11.40% of the renewable energy in the energy-mix system. However, the Earth's global surface temperature as stated in the Global Climate Summary is increasing up in July 2021 to 0.93°C above the average of 15.8°C where this is the highest temperature recorded [2]. This will lead to catastrophic climate issues if the world doesn't act now and depending too much on fossil fuel as the major energy production and consumption. The IPCC Report (2021) stated that climate change has been affecting every region on Earth in multiple ways and the changes will increase with additional warming

[3]. For instance, the highest temperature recorded this year has led to the largest wildfires in California, Colorado and the "black summer" of fires in eastern Australia [4]. Hence, immediate implementation of the low carbon energy system is necessary. This can be done by shifting the energy production from fossil fuels that releases an enormous amount of the CO_2 and other greenhouse gases to the low carbon and sustainable energy system to meet the needs of current generation without compromising the needs of the future generation.

In 2015, the treaty on climate change which is The Paris Agreement has been adopted internationally and, almost 200 countries have pledged to reach the main goal such that "to limit the global warming to well below 2°C, preferably to 1.5°C compared to pre-industrial level" [5][6]. In addition to that, the 2030 Agenda for Sustainable Development was introduced in the same year and focused on providing 17 Sustainable Development Goals (SDGs) plan of action for humanity, planet, and prosperity [7]. One of the SDGs is to develop affordable and clean energy (SDGs 7). According to IAEA, by harnessing nuclear energy in the future sustainable energy mix, SDGs 7 could be achieved and would help to reach another 16 goals indirectly as shown in Figure 1 [8].

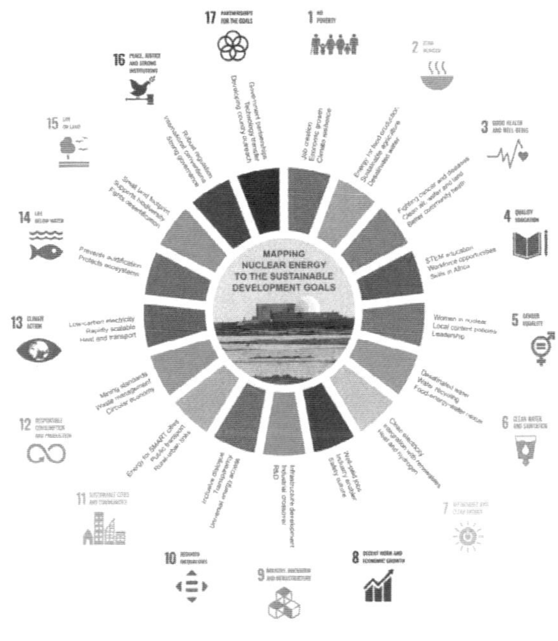

Figure 1. Nuclear energy is the key to the rest of SDGs [22].

In this review article, the focus area for the low carbon energy system discussed will be concentrating on the nuclear energy, specifically in the nuclear fuel materials subjected to its sustainability. The implementation of nuclear energy could help in reducing the CO_2 emission and greenhouse gases. However, issues in nuclear waste management is among the concern factors that limit users from harnessing the nuclear energy, due to the long-lived radioactive waste and lack of public acceptance [9]. Thus, this paper aims to review and discuss the fuel materials used in the conventional and advanced reactor technology related to its waste as well as radiotoxicity management, which can be elaborated in the following sections.

2. Nuclear fuel materials

Nuclear energy is generated via a fission process. In general, when a neutron is bombarded onto a heavy fissile nucleus, it will split into two daughter nuclei and release a huge amount of kinetic energy together with 2-3 fission neutrons as depicted in Figure 2. The fission process can be sustained by allowing another fission process to occur when the fission neutrons meet with another heavy fissile nuclei. This scenario is known as chain fission reaction and is used in current operating nuclear fission reactors to produce clean energy. The common types of heavy fissile nuclei for generating nuclear fission are Uranium (U) atom, Plutonium (Pu) atom and Thorium (Th) atom. These atoms can be divided into two isotope categories which are fissile isotopes and fertile isotopes as the fuel materials in the nuclear reactors. Fissile isotopes such as ^{235}U, ^{233}U, ^{239}Pu and ^{241}Pu required the thermal neutron to create the fission process. While ^{238}U and ^{232}Th are the example of fertile isotopes that required the fast neutron to convert the nucleus into fissile isotopes (^{239}Pu and ^{233}U) for fission process.

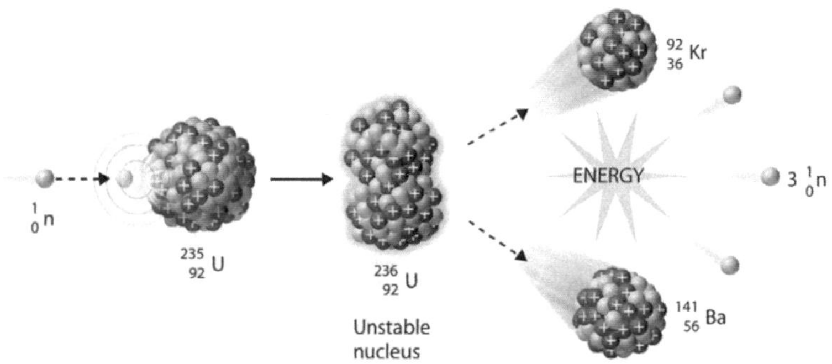

Figure 2. Fission reaction [10].

According to IAEA, the number of light water reactors (LWRs) in operation in the world, as on 31 Dec 2019, are 413 reactors, with 52 units are under construction while 73 reactors are planned to be implemented in future [11]. The current operating LWRs utilize the conventional enriched uranium oxide (UO_2) as the nuclear fuel materials. The composition of fissile materials in UO_2 is less than 5% which is commonly known as low enriched uranium (LEU). Besides LEU, mixed oxide (MOX) fuels were also used as the fuel materials in LWRs. The composition of MOX consists of 94-97% of UO_2 blended with 3-5% of the PuO_2 that has been extracted via reprocessing from the used nuclear fuels [12]. The extension on the reprocessing methods can be obtained in the next section.

The advancement of nuclear fuel materials has been conducted extensively to develope future nuclear energy with sustainable concepts. The Fukushima accidents has led to the improvement of the fuel materials in LWRs for advanced reactor applications via Accident Tolerance Fuels (ATF). The ATF should be designed to have the capability of performing better during normal operation as well as responding well when loss of coolant accident and reactivity-initiated accident to prevent reactor from meltdown as had happened in Fukushima plant [13]. For instance, fully ceramic microencapsulated fuel (FCM) with tri-structural isotropic (TRISO) particles, uranium carbide (UC), uranium nitride (UN), uranium dicarbide (UC_2) are among the potential candidates of the advanced nuclear fuel materials for advanced reactor technology besides LWRs types [14][15]. In addition, more researchers are also interested in investigating the higher composition of uranium enrichment between 5% to 19.75% called high-assay low enriched uranium (HALEU) as the advanced nuclear fuel materials. A study done by [16] suggested that the reactor performances and safety analysis of LWRs when using HALEU composition in the fuel concluded that no significant neutronic or reactor safety interruption could be observed, however further exploring the potential of HALEU in LWRs was recommended. Besides, the study also claimed that the waste volume per unit energy generated can be reduced but required additional natural resources due to higher enrichment of the ^{235}U isotopes. HALEU also may be needed in future advanced reactor technology for improving the fuel utilization and helps in plant economics [17]. Furthermore, the advancement of nuclear reactor technology such as long-term reactor operations, advances nuclear fuel materials, improvement on the reactor design as well as minimizing the waste volume and cost-effectiveness makes the nuclear energy are more feasible and sustainable to the environment, economic and social [18].

3. Nuclear spent fuel volume

Besides the radiation hazard and public acceptance issues, the volume of the used nuclear fuel and waste produced in the nuclear power plant causing this type of low carbon energy system is not reliable for sustainability. In addition, the long-lived radioactive and the high heat load in the spent fuels contributes to the development of an intensive nuclear waste management, deep geological repository and poses a great danger to humankind and the environment. Current operating nuclear reactors are of the water-cooled reactors, commonly known as LWR. The spent fuel produced from these reactors consists of major actinides, minor actinides, and short-lived fission products. Approximately 0.4% of these cumulative inventories are classified as high-level waste (HLW) with half-life more than 2.0 x10^5 years [19]. According to [20], around 150 tonnes of the HLW and fission products were stored permanently in the underground nuclear waste facilities worldwide in 2006. And the trend of the spent fuel being unloaded from the reactors will increase by 11,500 tHM in the year of 2010 annually as depicted in Figure 3. Thus, the volume of the HLW waste also will keep on increasing and eventually, making nuclear energy unsustainable, although this energy is clean and produces zero CO_2 emission during energy generation.

Hence, to overcome this issues, various studies have been conducted with the aim to reduce the waste volume by extracting and reprocessing the useful actinides from the spent fuels for the peaceful use rather than treating it as HLW directly and disposing it permanently. The extraction of useful actinides such as Pu could lead to the proliferation threat. Due to this issue, in 1953, the Atoms for Peace program for regulating the uses of nuclear technology for peaceful purposes was announced [21]. In 1968, the Treaty on the Non-Proliferation of Nuclear Weapons (NPT) was introduced to all countries with its objective to prevent the spread of nuclear weapons and weapons technology as well as to promote the cooperation in the nuclear energy for peaceful uses [22]. The efforts made by almost 191 States that joined to follow the NPT has promoted the advancement of nuclear technology to emerge with other's fields without worrying about the proliferation threat.

As stated in IAEA report, the cumulative of the spent fuel generated around the world in the beginning of 2004 was 268 000 tHM, where 90 000 tHM had been repossessed [23]. Based on Figure 3, by conducting the reprocessing methods, the amount of the spent fuels that needed to be stored will decrease annually and helps this clean energy production to be among the sustainable energy for achieving the sustainable development goals (SDGs).

The current strategy to reduce the waste volume can be done by partitioning and transmutation (P&T) process. This process could help in reducing the waste volume for peaceful uses. For instance, the process works in two parts. The first part is partitioning where it is carried out by separating the actinides, including Pu and U in the spent fuels using the pyrochemical technologies to produce the new nuclear fuel from the waste.

Next, the separated actinides will be burned and transmuted into the short-lived radioactive materials in the transmutation process where it can be carried out in any types of nuclear reactor theoretically, but preferable to be conducted in fast breeder reactors (FBR) and accelerator driven system (ADS) [24-26]. The pyrochemical technologies has been conducted extensively in many studies worldwide as reported by OECD in their status report [27]. In addition, the study conducted by [28] to investigate the significant capability of P&T methods for decreasing the waste volume was proven where the number of vitrified waste packages can be reduced to 1/4 which also lessens the waste facility footprint to 1/100 effectively. In [29] work, the comparison of waste produces with the P&T and non-P&T methods showed substantial waste package and footprint reduction by 54% incorporating the proposed 'high-waste-loading glass and Cold Crucible Induction Melter (CCIM)' waste case as stated in the study. Since this paper is only focusing on reviewing the suitability of nuclear fuel materials for sustainability, thus the details on the technical parts of the P&T and the pyrochemical technologies will not be discussed.

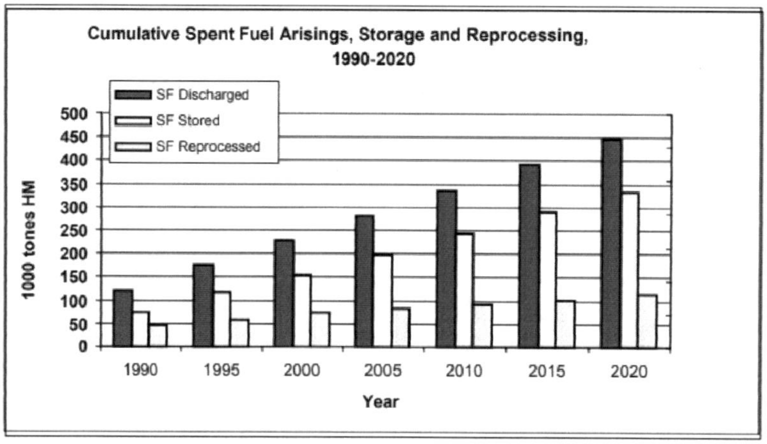

Figure 3. Total amount of spent fuels to be stored, discharged, and reprocessed [20].

4. Radiotoxicity

The parameter to determine the toxicity released by the nuclear spent fuel is known as radiotoxicity. This parameter is an essential information required in managing the waste disposal facilities either in temporary storage facilities or geological repositories for long-term disposal. It is well noted that radiotoxicity is the result of the decay actinides and fission products which mainly depends on the fuel materials such as UO_2, MOX or Th fuel. Figure 4 shows the trend of the radiotoxicity over the years after reactor's shutdowns based on U and Th fuels. For the radioactivity during and after the reactor operation, in comparison with radioactivity of the spent fuel during its subsequent handling, several things must be taken into consideration. Even during the operation, it is very difficult to determine the exact number of different radionuclides within the fuel. This is because many radionuclides have very short lives, with behaviour that instantly disappears after the reactor shutdown.

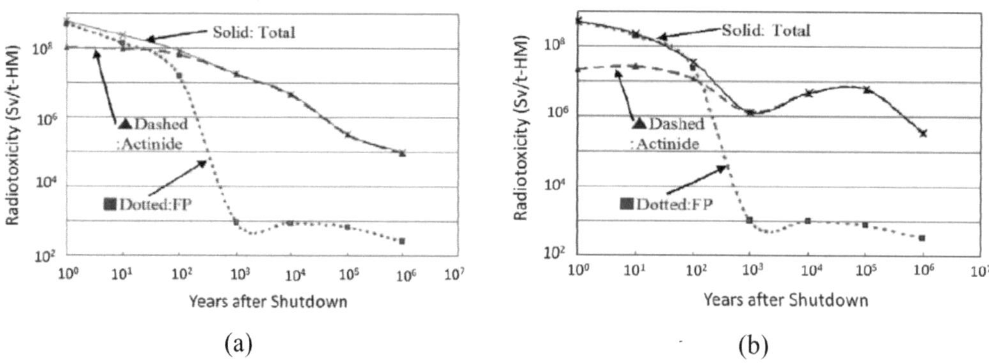

Figure 4. Component of radiotoxicity for the (a) U fuels and (b) Th fuels [30].

Radiotoxicity is mainly influenced by short-lived nuclides, including iodine. Majority very short half-lives that only contribute to a small percentage of the overall radiotoxicity after one month. Some comparison studies have been conducted to investigate the radiotoxicity among the fuel material types. For example, reactors with Th fuels produce less Pu than the U fuels which result in lesser radiotoxicity, but a daughter nuclide of ^{233}U in the Th reactors causes higher radiotoxicity during long-term storage than the U core [30]. For the following Pressurized Water Reactor (PWR) cores, the ORIGEN2 code [31] was used to estimate the amount of radiotoxicity. For fuels irradiated to 45 GWd/t, a constant rated power density of 39.2 MW/t was used, and radiotoxicity was analysed for time periods up to 1 million years.

- U-PWR-core: 4% ^{235}U + 96% ^{238}U, ^{220}U-Library.
- Th-PWR-core: 4% ^{233}U + 96% ^{232}Th, ^{214}Th-Library.

Equation (1) was used to translate the radioactivity calculated by ORIGEN2 to radiotoxicity. Note that ORIGEN2 gives the radioactivity of a tonne of HM (initial Heavy Metal) in 'Curie' units, which must be converted to 'Becquerel' units.

$$Radiotoxicity\ [Bq/t - HM] = Radioactivity\ [Bq/t - HM]\ x\ DCF\ [Sv/Bq] \quad (1)$$

When a nuclide is injected into a human body, it generates internal radiation exposure, which is measured by radiotoxicity. It is determined by the type of radiation, its energy, and its influence on the human body, which is described as a dose conversion factor (DCF). The majority of DCFs may be found in numerous references, such as [32] and [33]. DCFs for several other nuclides can also be found in [34] and [35], or authors can make assumptions for them. A comparison of radiotoxicities calculated for the U and Th cores in PWR is shown in Figure 5. Up to 1000 years, the Th core has a 10–90 percent lower radiotoxicity than the U core; beyond that, the tendency reverses. Around 0.1 million years, there is a peak in radiotoxicity. The decay heat curves are found to be similar to these curves. Table 1 shows the numerical values of the data in Figure 5.

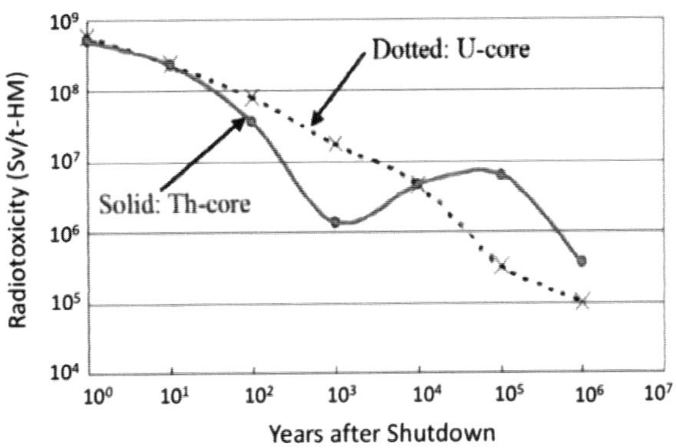

Figure 5. Radiotoxicity for U core and Th core.

Table 1. Radiotoxicity for U core and Th core (Sv/t-HM).

Year	*1*	*10*	*100*	*1.00E+03*	*1.00E+04*	*1.00E+05*	*1.00E+06*
U Core	5.88E+08	2.36E+08	8.18E+07	1.77E+07	4.59E+06	3.30E+05	9.80E+04
Th Core	5.18E+08	2.32E+08	3.57E+07	1.32E+06	4.65E+06	6.16E+06	3.51E+05
%	-12.0	-1.9	-56.4	-92.5	1.3	1766.9	258.5

^{239}Pu, which contributes about half of the overall radiotoxicity in the U core during long-term storage (e.g., 10 000 to 100 000 years), is the most important contributor of radiotoxicity during long-term storage (e.g., 10 000 to 100 000 years). ^{229}Th is the largest contributor of radiotoxicity in the Th core during long-term storage (e.g., from 10,000 to 100,000 years), accounting for over half of the total radiotoxicity. ^{233}U has a daughter nuclide, ^{229}Th (half-life: 159 200 year). Thus, the Th core produces less Pu than the U core, resulting in lesser radiotoxicity from Pu, but a daughter nuclide of ^{233}U in the Th core causes higher radiotoxicity during long-term storage than the U core.

In [36] stated that the cumulative radiotoxicity of the actinides in the U-Pu fuels and Th-U fuels were 2.5 times higher and 3.5 times lower than the U fuels respectively. Effort has been made to reduce the toxicity of the spent fuel which can be carried out by recycling the actinides and fission products in the spent fuel to decrease the nuclear waste volume via P&T process. The separation and transmutation of the minor actinides may contribute to shortening the long-term decay heat and radiotoxicity inventories. However, in [28] works claimed that only ^{90}Sr and ^{137}Cs of the minor actinides produced significant depletion on the radiotoxicity using P&T process in the deuteron accelerator. In [37] stated that the transmutation of Pu by irradiating all the MOX fuels in current operating LWRs in the next 50 years would require a long period of time for radiotoxicity reduction.

Chemical separation of Plutonium, Americium, and Curium before long-term controlled storage is recommended for Uranium and Plutonium spent fuels. For significant conversion of ^{241}Pu into ^{241}Am, Americium should be separated after 50-70 years of storage. ^{244}Cm decays almost completely after 100 years. Extracted americium should be utilised for transmutation and plutonium should be reused. Separation of actinides is also effective in lowering the heat of decay.

^{232}U determines the vast majority of radiotoxicity in Thorium spent fuel. It is self-evident that repeated usage of Thorium fuel will result in an increase in radiotoxicity. Additional deep burn-out (transmutation) of Uranium fraction comprising both ^{233}U and ^{232}U is required for one-fold utilization of thorium fuel with deep ^{233}U burnup. The extraction and transmutation of Plutonium fractions can reduce radiotoxicity by several orders of magnitude (^{238}Pu). Because ^{228}Th decays virtually completely in 10 years with its short-lived daughter nuclides, transmutation of ^{228}Th - daughter nuclide of ^{232}U is not required.

5. Discussion

From the literature review, the nuclear fuel materials improvement in advancing the reactor technology to make it more feasible and sustainable has been conducted extensively. Align with the Paris Agreement and the 2030 Agenda on Sustainable Development, the application of nuclear energy could help in reducing the CO_2 and greenhouse gases emission. The current contribution of nuclear energy for electricity generation has been reported to avoid nearly 20% of the global CO_2 emission from 1971-2018 [38]. Whereas the greenhouse gases emissions from nuclear energy are among the lowest compared to other energy sources as depicted in Figure 6 [39].

However, to fully utilize nuclear energy by replacing the current fossil fuel-fired power plant is not feasible as it still has the risk potential from the waste and large releases of radioactive materials. The improvement in fossil fuel-fired power plant also has been made to reduce the CO_2 emission and greenhouse gases via Carbon Capture and Storage (CCS). In [40] report, the usage of CCS has captured almost 40 Mt of CO_2 in 2020 and expected to capture more in 2050 around 5,635 Mtpa if the application of CCS is speed up globally. In addition, the net installed renewable energy is expected to grow. As reported in [41], the addition of wind and hydropower are increasing drastically, accounting for almost 90% of the total global power capacity, while the application of the solar photovoltaic remain stable and on demand. Hence, the authority especially the policymakers need to come out with the solution on the future energy mix strategies by including all these three types of energy generation from nuclear energy (advanced reactor technology), fossil fuel-fired power plant with CCS and renewable energy (wind, hydropower, solar) to solve the climate issues as well as saving the environmental and human from the consequences of insecure energy source. In addition, by installing the CCS facilities powered by nuclear energy for capturing the CO_2 also could resolve the global warming issues to achieve sustainable development.

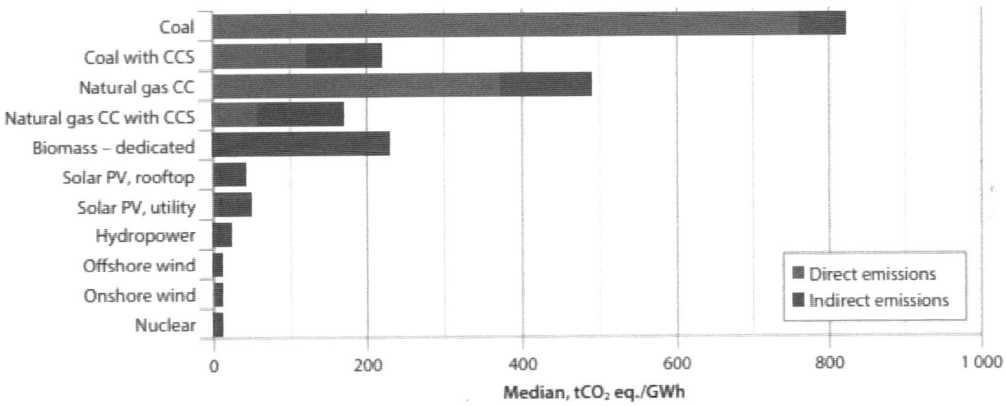

Figure 6. Greenhouse gases emissions comparison [35].

6. Conclusion

This paper outlined the advantages of having nuclear energy as the key to overcome climate change issues by focusing on sustainability aspects of the nuclear fuel materials. The conventional fuel types for LWRs and advanced fuel materials were discussed as well as the P&T and the pyrochemical technologies in reducing the radiotoxicity, waste volume and heat load of spent fuels. It can be concluded that by having the advanced nuclear fuel materials with advanced reactor technology in the energy mix strategies can help in reducing the CO_2 gases emission and greenhouse effects and, providing the world with effective solutions for energy security besides depending on the fossil-fuel fired power plants for continuous energy production.

7. References

[1] Hannah R and Max R 2020 *Energy*. Published online at OurWorldInData.org.
[2] Rebecca L 2021 Earth's hottest month was record hot in 2021.
[3] IPCC 2021 IPCC press release AR6 *Climate Change 2013 - The Physical Science Basis,* 1–6.
[4] Isabelle G 2021 After the turbulent year of 2020, BBC Future takes stock on the state of the climate at the beginning of 2021.
[5] Delbeke J, Runge-Metzger A, Slingenberg Y and Werksman J 2019 The Paris agreement. *Towards a Climate-Neutral Europe: Curbing the Trend* 24–45.

[6] Green C 2015 A "fair and ambitious" climate agreement is not nearly enough: Paris 2015 take heed! *Environ.Res. Letters*, **10**(10) 9–11

[7] United Nation 2015 Transforming our world: the 2030 Agenda for Sustainable Development: https://www.un.org/ga/search/view_doc.asp?symbol=A/RES/70/1&Lang=E

[8] The United Nations Economic Commission for Europe 2021 Application of the United Nations Framework Classification for Resources and the United Nations Resource Management System: Use of Nuclear Fuel Resources for Sustainable Development-Entry Pathways.

[9] R Haas Eds the Technological and Economic Future of Nuclear Power, Energiepolitik und Klimaschutz *Energy Policy and Climate Protection.*

[10] John R 2014 Nuclear fission *AccessScience* 1-13

[11] IAEA 2020 Nuclear Power Reactors in the World *Atomic Energy* 1-81.

[12] Carbajo J, Yoder L, Popov G and Ivanov K 2001 A review of the thermophysical properties of MOX and UO2 fuels. *J. Nucl. Mater.* **299**(3) 181–198

[13] Was S, Petti D, Ukai S and Zinkle S 2019 Materials for future nuclear energy systems *J. Nucl. Mater.* **527** 151837

[14] Worrall A and Snead A 2015 Technology Implementation Plan: Fully Ceramic Microencapsulated Fuel for Commercial Light Water Reactor Application *ORNL/TM-2015* 220.

[15] Grande L, Villamere B, Allison L, Mikhael S, Rodriguez-Prado A, and Pioro I 2011 Thermal aspects of uranium carbide and uranium dicarbide fuels in supercritical water-cooled nuclear reactors. *J. Eng. for Gas Turbines and Power* **133**(2) 1–7.

[16] Burns R, Hernandez R, Terrani A, Nelson T and Brown R 2020 Reactor and fuel cycle performance of light water reactor fuel with 235U enrichments above 5%. *Annals of Nuclear Energy* **142** 107423

[17] Testoni R, Bersano A and Segantin S 2021 Review of nuclear microreactors: Status, potentialities and challenges. *Prog. in Nucl. Energy* **138** 103822.

[18] Matthew F 2020 Advanced Reactors Help Pave the Way for Nuclear Power to Meet Climate Goals *IAEA Bulletin* **61**-3

[19] Al Qaaod A, Shahbunder H, Refeat R. M, Amin A and El-Kameesy U 2018 Transmutation performance of uniform and nonuniform distributions of plutonium and minor actinides in TRIGA Mark II ADS reactor *Ann. of Nucl.Energy* **121** 101–107.

[20] International Atomic Energy Agency 2009 Status of Minor Actinide Fuel Development. *IAEA Nuclear Energy Series* 1–95.

[21] William C Atoms for Peace Program *Encyclo. of Sci. Technol. and Ethics.*

[22] Circular I 2021 Nuclear non-proliferation *Routledge Handbook of Security Studies* 395–406

[23] Lovasic Z 2008 Spent Fuel Reprocessing Options. *Nucl. Fuel Cycle and Mater. Sect.* 151

[24] Ackerman J 1991 Chemical Basis for Pyrochemical Reprocessing of Nuclear Fuel *Industr. and Eng. Chem. Res.* **30** 141–145

[25] Fujita R and Akai Y. 1998 Development of transuranium elements recovery from high-level radioactive liquid waste. *J. Alloys and Comp.* 271–273

[26] Kooyman T. 2021 Current state of partitioning and transmutation studies for advanced nuclear fuel cycles. *Ann. of Nucl. Energy* **157** 108239

[27] OECD 2004 Pyrochemical separations in nuclear applications **5427**

[28] Fukaya Y, Ueta S, Yamamoto T, Chikazawa Y and Yan L 2021 Toxicity Reduction with Total Volume Control in Nuclear Waste *Nucl. Technol.* **00** 1–12

[29] Fukaya Y 2020 Optimization of light water reactor High Level Waste disposal scenario in the situation of delayed reprocessing with existing and demonstrated technology. *Ann. of Nucl. Energy* **144** 107503
[30] Gerasimov S Zaritskaya T Kiselev V and Myrtsymova A 2001 *9 Int., Conf., on Nucl. Eng.* (France)
[31] Croff A G 1980 *Nucl. Fuel Cycle And Fuel Mater.* **11**(21)
[32] ICRP 2012 *Ann. ICRP*, **119** 130.
[33] Kawai K, Endo A and Noguchi H 2002 *Radiat. Protect. And Dosimetry (S61)* **37**(1)
[34] Sowby F D 1981 *Ann. of the ICRP* **6**(1), 1.
[35] Wada S, Hiraiwa K, Yoshioka K, Kimura R, Sakurai S and Sugita T 2020 *Nucl. Fuel Cycle and Mater. Sect.* **51**(1).
[36] Gerasimov S Bergelson R Zaritskaya S and Tikhomirov V 2002 Comparison of decay heat power of uranium, plutonium, and thorium spent fuel at long-term storage. *10th Int. Conf. on Nucl. Eng.* **4** (Arlington, Virginia, USA) pp 481–484.
[37] Trellue R 2003 Reduction of the radiotoxicity of spent nuclear fuel using a two-tiered system comprising light water reactors and accelerator-driven systems.
[38] OECD 2019 Nuclear Power in a Clean Energy System
[39] Organisation for Economic Co-operation and Development 2012 The role of nuclear energy in a low-carbon energy future (OECD Publishing).
[40] Institute C 2017 Global costs of carbon capture and storage (Global CCS Institute).
[41] Muellner N, Arnold N, Gufler K, Kromp W, Renneberg W and Liebert W 2021 Nuclear energy - The solution to climate change? *Energy Policy* **155** 112363.

Assessment of rare earth and actinides (U and Th) elements in soil samples from Kapar industrial area, Selangor

M S Elias[1], M A Azman[1], J A D Daung[2], A Hashim[2], Z Laili[1], S A Omar[1] and S Shukor[1]

[1]Analytical Chemistry Application Group (ACA), Waste and Environmental Technology Division, Malaysian Nuclear Agency, Bangi, 43000 Kajang, Selangor, Malaysia.
[2]Environmental Tracer Application Group (E-TAG), Waste and Environmental Technology Division, Malaysian Nuclear Agency, Bangi, 43000 Kajang, Selangor, Malaysia.

mdsuhaimi@nuclearmalaysia.gov.my

Abstract. The study of rare earth and actinide elements have received attention nowadays due to rapid development in the extraction of the rare earth industry in Malaysia. Soil samples (18 areas) collected from the Kapar industrial area were analysed by using the neutron activation analysis (NAA) technique for the determination of rare earth elements and actinides (U, Th) concentration. The concentration of rare earth element such as La, Ce, Nd, Sm, Eu, Tb, Dy, Yb and Lu were ranged from 14.7 – 61.4, 30.6 – 132.8, 11.7 – 44.0, 2.12 – 9.60, 0.37 – 1.43, 0.03 – 1.57, 2.20 – 7.45, 0.28 – 0.81, and 1.99 – 4.15 mg/kg, respectively. The concentrations of uranium and thorium were a range of < 0.1 – 6.74 and 12.6 – 30.1 mg/kg, respectively. The results of geo-accumulation indicate the all-rare earth elements can be categorised as uncontaminated in all areas with an I_{geo} value below 0. The concentration of U can be categorised as uncontaminated to moderately contaminated (I_{geo} class 1) with I_{geo} values of 0 to 1 in the area of D, F, H, I, L, and N, whilst other areas can be categorised as uncontaminated with I_{geo} value < 0 (I_{geo} class 0). The thorium concentration in B, L, and N areas can be categorised as moderately contaminated with I_{geo} value 1 to 2 (I_{geo} class 2), whilst most of the area can be categorised as uncontaminated to moderately contaminated (I_{geo} values 0 to 1). The sources of Th contamination most likely originated from the anthropogenic and geogenic processes.

1. Introduction

In Malaysia, the pollution issues of rare earth elements (REEs) and actinide elements (U and Th) have become the attention of authorities from the government agencies, non-governmental organisations (NGOs), researchers, and members of the public. Rapid development in the rare earth industry in Malaysia related to REEs activities such as mining, mineral and rare earth processing, recycling and recovery of by-products, waste treatment of rare earth, and final disposal is also an issue. The U and Th elements become attention due to health hazards related to the radiation level. However, the effect of REEs concentration is still uncertain because the unknown threshold standard level of REEs will have an adverse effect on human health and other organisms.

Soil contamination may be increased due to continuous input from various human activities, such as industrial, agriculture, mining, logging activities and residential, highway, and road development.

Besides that, soil contamination can also originate from geogenic processes such as landslides, terrestrial runoff, and land erosion. Kapar area is one of the industrial areas in Selangor state of Malaysia. Many activities in the Kapar area include electrical, cable industry, aluminium smelter, glove, steel manufacturing, coating, plastic, and wood industries. Besides that, Kapar area was located the coal power plant, which could be contributed to the pollution. The objectives of the study in Kapar industrial area are to identify the level and distribution of rare earth and actinides (U and Th) elements.

Study of rare earth, U, Th, heavy metal and trace element of Gebeng industrial area, Manjung and central of Perak, Sabah, and Sarawak area in soil has been done for the purpose to identify the distribution, concentration level, and status of soil contamination[1–3]. Generally, the contamination level (I_{geo}) of RREs in the soil of the Gebeng industrial area can be classified as uncontaminated to moderately contaminated with I_{geo} values was ranged from < 0 to 1.6, whilst uranium (U) can be categorised as uncontaminated (I_{geo} value < 0) and thorium (Th) was categorised as uncontaminated to moderately contaminated (I_{geo} value $< 0 - 0.9$). The REEs contamination in soil of Manjung and central of Perak (0-10 cm depth) can be classified as uncontaminated to severely contaminated with I_{geo} value was ranged from less than 0 to 4.38. The uranium can be categorised as uncontaminated to moderately contaminated (I_{geo} value $< 0 - 2.60$), whilst the thorium can be categorised as uncontaminated to slightly contaminated (I_{geo} value $< 0 - 1.98$)[2]. The sources of soil contamination in the Manjung and central of Perak was originated from widely used fertiliser in agriculture activities which contains Dy, Lu, Tb, and U [2].

2. Methodology

2.1. Sampling of soil samples from Kapar industrial area.

Soil samples (15 cm depth) were collected from 18 areas of the Kapar Industrial area using an auger and transferred into a polyethylene bag (1 kg). Soil samples were then transported to the laboratory. At the laboratory, the soil samples were then dried at 60°C in an oven and ground using agate mortar. The powdered soil samples were kept in polyethylene bottles for analysis. The sampling areas of soil and the description of activities at the sampling site of Kapar areas were depicted in Figure 1 and Table 1, respectively.

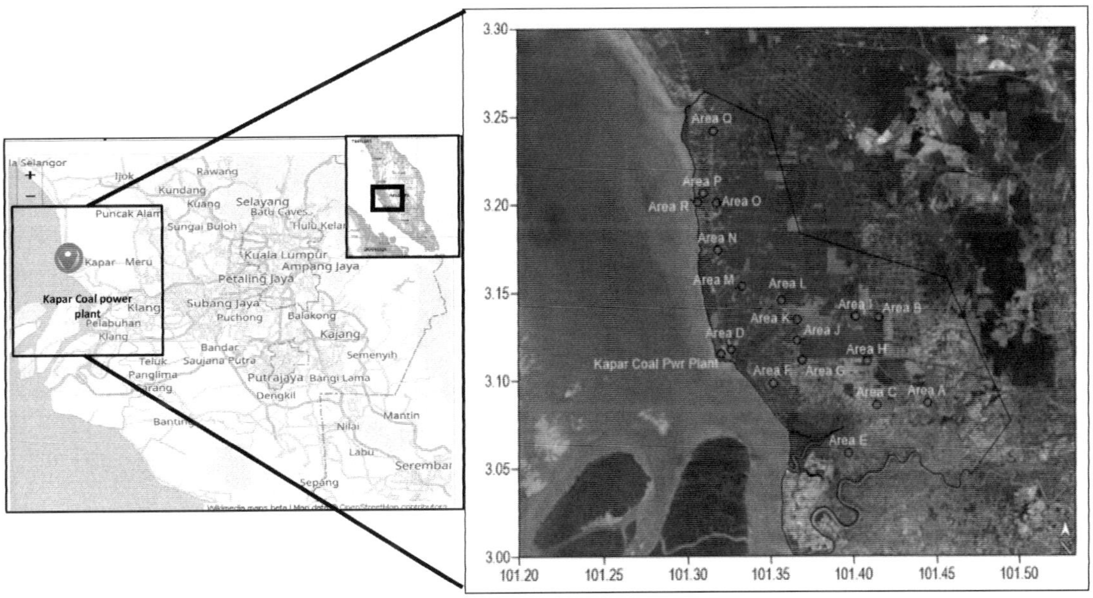

Figure 1. Sampling locations of soil sample collected from Kapar Industrial area.

Table 1. The description of activities in the sampling site.

Location	Longitude	Latitude	Activities description
Area A	101.444050	3.086883	Residential area
Area B	101.414833	3.135467	Meru South Industrial Park – Electrical, cable industry, aluminium smelter, glove and steel industry
Area C	101.413933	3.086017	Sg. Kapar Industrial Park – automotive, plastic, glove, wood, steel, costing industries.
Area H	101.408100	3.110433	
Area D	101.325917	3.117333	Close to Kapar coal power plant, and agriculture area
Area E	101.396900	3.058667	Agriculture area, and residential area
Area F	101.351550	3.098500	Residential area
Area G	101.368967	3.111217	Residential area, and shop lot
Area I	101.401333	3.136033	Kapar Industry Park – manufacturing, coating, plastic, and wood industry.
Area J	101.365667	3.122550	
Area K	101.365683	3.134350	
Area L	101.356283	3.145483	Residential area, and small agriculture area
Area M	101.332883	3.153767	Residential area, and small agriculture area
Area N	101.318383	3.173833	Tambak Jaya - Food industry, manufacturing, wood, furniture, plastic and coating industry
Area O	101.317650	3.200883	Agriculture area, and residential area
Area P	101.309333	3.206017	Agriculture area, and residential area
Area Q	101.315367	3.241900	Residential area
Area R	101.306417	3.201217	Agriculture area, and residential area

2.2 *Analysis of soil samples by using Neutron Activation Analysis (NAA) technique*

Approximately 0.20 g of powdered soil samples were weighed using an analytical balance and transferred into small polyethylene vials (Ø = 1 cm × H = 3 cm) and sealed prior to the irradiation process. The duplicate soil samples, blank, mix standard and standard reference materials (SRMs) (IAEA-SL-1, SRM 2709a-San Joaquin Soil) were irradiated in the PUSPATI TRIGA Mark II Reactor at the Malaysian Nuclear Agency for 6 hours at 750 kW, with a thermal flux of 4.0 x 10^{12} n.cm^{-2}s^{-1} (RR-Rotary rack)[4–7]. The cooling times of soil samples, blank, mix standard, and SRM were varied from 2 to 4 days for the first counting and 3 to 4 weeks for the second counting. The counting process for the analysis of long half-life radionuclides of rare earth elements (La, Ce, Nd, Sm, Eu, Tb, Dy, Lu, and Yb) and actinides elements (U and Th) in soil samples, blank and SRM were performed for one hour each by using high-resolution hyper-pure germanium (HpGe) gamma detector. A comparative method was applied to measure the elemental concentrations in the sample. The comparative method was based on the comparison between the net count of the samples and the weight of the standard to the net count of the standard and weight of the sample according to the Eq. (1) [4,8,9]and the data were reported in dry weight (d.w.).

$$C_{smp} = C_{std} \times (W_{std} \times N_{smp})/(W_{smp} \times N_{std}) \quad (1)$$

where: N_{smp} and N_{std} are the net counts of the peak area of the sample and standard, respectively; C_{smp} and C_{std} are the concentration of the sample and standard, respectively; W_{smp} and W_{std} are the weight of the sample and standard, respectively.

2.3 *Contour map*

All contour maps of REEs, U and Th concentrations were done using Surfer software (version 11). Kriging interpolations were conducted to visually examine the contour map of the REEs, U and Th concentration distribution in the soil of the Kapar industrial area.

3. Results and discussion

3.1 The concentration of rare earth elements (REEs), U and Th in soil sample

Soil samples were collected in eighteen selected areas in Kapar industrial for the analysis of elemental pollution using the Neutron activation analysis (NAA) technique. The concentrations of rare earth, uranium, and thorium were tabulated in Table 3.1. The concentration of rare earth element such as La, Ce, Nd, Sm, Eu, Tb, Dy, Yb, and Lu were ranged from 14.7 – 61.4, 30.6 – 132.8, 11.7 – 44.0, 2.12 – 9.60, 0.37 – 1.43, 0.03 – 1.57, 2.20 – 7.45, 0.28 – 0.81, and 1.99 – 4.15 mg/kg, respectively whilst the concentrations of uranium and thorium (actinides) elements were range of < 0.1 – 6.74 and 12.6 – 30.1 mg/kg, respectively. Average concentrations of all elements of rare earth showed lower than igneous granitic rock- IGR crust values. However, the average concentration of U (4.13 mg/kg) and Th (20.3 mg/k) showed slightly higher than the igneous granitic rock- IGR crust values as tabulated in Table 2.

All elements of rare earth showed obviously higher in N area compared to other areas. The activities of the N area are related to the food industry, manufacturing, wood, furniture, plastic, and coating industries. The concentration of uranium (6.74 mg/kg) showed higher in the I area which is located at Kapar Industry Park and most activities related to manufacturing, coating, plastic, and wood industries. The B area is located at Meru South Industrial Park showed a higher thorium concentration (30.1 mg/kg) and was mainly related to industries of electrical, cable, aluminium smelter, glove, and steel.

Table 2. The concentration (mg/kg) of rare earth element in soil of Kapar Industrial area.

Location	La	Ce	Nd	Sm	Eu	Tb	Dy	Lu	Yb	U	Th
Area A	26.2	56.2	22.9	3.82	0.68	0.94	3.66	0.57	3.12	3.27	23.9
Area B	24.9	39.9	20.3	2.12	0.37	0.53	2.20	0.59	3.07	3.92	30.1
Area C	14.7	30.6	11.7	2.38	0.40	0.55	3.83	0.41	2.30	3.21	13.8
Area D	38.6	89.4	31.3	6.02	0.79	0.97	6.83	0.46	2.73	4.58	20.4
Area E	20.6	74.2	21.1	3.48	0.60	0.75	4.42	0.44	2.42	3.83	16.9
Area F	34.7	78.5	37.5	5.55	0.73	0.92	5.43	0.43	2.85	4.90	22.4
Area G	61.3	42.7	17.4	2.55	0.56	0.63	3.52	0.42	2.15	2.77	21.3
Area H	31.8	80.3	30.8	4.93	0.67	0.76	3.62	0.28	2.05	5.18	20.5
Area I	24.2	54.4	20.4	5.52	0.67	0.03	4.37	0.39	3.21	6.74	19.6
Area J	16.7	37.7	14.4	3.13	0.44	0.61	3.25	0.38	2.30	<0.1	19.9
Area K	17.2	40.6	12.6	3.48	0.53	0.76	3.13	0.36	2.33	<0.1	19.0
Area L	17.5	37.8	15.4	4.25	0.71	1.01	4.50	0.61	3.63	5.08	26.5
Area M	25.5	60.6	28.9	4.82	0.65	0.72	3.40	0.29	1.99	3.25	12.6
Area N	61.4	133	44.0	9.60	1.43	1.57	7.45	0.81	4.15	6.35	28.2
Area O	32.8	70.5	24.1	5.07	0.78	0.72	4.13	0.49	2.41	2.89	18.0
Area P	46.8	68.3	35.8	5.67	0.74	0.03	2.68	0.33	3.25	3.91	16.7
Area Q	37.1	58.4	25.4	4.57	0.61	0.03	2.88	0.30	2.81	3.18	14.4
Area R	27.3	63.1	15.8	3.62	0.58	0.80	4.71	0.31	2.46	3.00	20.3
Average	31.1	62.0	23.9	4.48	0.66	0.69	4.11	0.44	2.74	4.13	20.3
IGR crust*	45.0	81.0	33.0	8.80	1.40	1.40	6.30	1.10	3.50	3.00	8.50

*Igneous granitic rock- IGR crust value published by Turekian and Wedepohl, 1961.

3.2 Geo-accumulation (I_{geo}) index

The geo-accumulation index (I_{geo}) was originally introduced by Müller (1969)[10]. The I_{geo} was used to define the degree of anthropogenic pollution by comparing the current concentration status of elemental over the pre-industrial levels of pollution in soil or sediment [4,11–15]. In Malaysia, the lack of the pre-industrial (background) data value of elements concentration acts as a reference crust value. Several authors published the reference crust value of elements such as the upper continental crust – 1 (UCC – 1)[16], mud from Queensland (MUQ)[17], igneous granitic rock (IGR)[18] and continental crust average (CCA)[19] for the comparison. In this study, rare earth element, uranium and thorium contamination in the soil of Kapar industrial area was compared to the igneous granitic rock-ICR crust values for the calculation of geo-accumulation index as described by Elias, 2021[20]. The geo-accumulation index can be calculated using Eq. (2).

$$I_{geo} = log_2\left(\frac{C_n}{1.5 \times B_n}\right) \quad . \tag{2}$$

where C_n is the measured concentration of interest element, B_n is the background value of interest reference element concentration from crust value (igneous granitic rock- IGR crust) [18] and 1.5 is a correction factor for the variation of the background values due to lithogenic effects in the soil. I_{geo} can be classified into seven classifications to determine the level of soil pollution. The soil pollution classification are as follows: I_{geo} value < 0 = uncontaminated (I_{geo} class 0), I_{geo} value > 0 to 1 = uncontaminated to moderately contaminated (I_{geo} class 1), I_{geo} value > 1 to 2 = moderately contaminated (I_{geo} class 2), I_{geo} value > 2 to 3 = moderately to strongly contaminated (I_{geo} class 3), I_{geo} value > 3 to 4 = strongly contaminated (I_{geo} class 4), I_{geo} value 4 to 5 = strongly to extremely contaminated (I_{geo} class 5) and I_{geo} value > 5 = extremely contaminated (I_{geo} class 6)[11–15,21].

The status of the geo-accumulation index, the jitter and whisker plot of I_{geo} value for the RREs, U, and Th contamination in the soil of Kapar industrial area was depicted in Figure 3.1. The horizontal line showed the median value in the box plot, whilst the lower border of the horizontal line illustrates the 25 percentiles value and 75 percentiles value are the top borders of the horizontal line of the box.

I_{geo} values and I_{geo} class of the REEs, U and Th in soil sample of Kapar industrial area were tabulated in Table 3. The I_{geo} values of all REEs shown below 0 indicate uncontaminated (I_{geo} class 0) except for the Nd element in the area of N with I_{geo} value of 0.13 (I_{geo} class 1). The concentration of REEs in the N and other areas is still considered uncontaminated (no instruction from industries activities) based on the geo-accumulation index, as shown in Figure 3.1. The I_{geo} value of the U in the area of D, F, H, I, L, and N were 0.03, 0.12, 0.20, 0.58, 0.18, and 0.50, respectively (Table 3.2). Those areas can be categorised as uncontaminated to moderately contaminated with I_{geo} class 1, while other areas were categorised as uncontaminated with I_{geo} value < 0 (I_{geo} class 0). The area of B, L, and N can be categorised as moderately contaminated with I_{geo} values = 1 to 2 (I_{geo} class 2) whilst other areas can be categorised as uncontaminated to moderately contaminated with I_{geo} value = 0 to 1 (I_{geo} class 1) for the contamination of thorium element. Precaution needs to be taken by the industries to prevent the release of elemental contamination into the soil by regularly monitoring the release of dust from the chimney and retention pond, especially in the areas that were categorised as moderately contaminated (I_{geo} class 2). The government agencies and authorities should regularly monitor the industries to follow the Environment Quality Act 1974.

IOP Conf. Series: Materials Science and Engineering 1231 (2022) 012017 doi:10.1088/1757-899X/1231/1/012017

Table 3. I_{geo} values and I_{geo} index of the REEs, U, and Th in soil sample of Kapar Industrial Area.

Location	La	Ce	Nd	Sm	Eu	Tb	Dy	Lu	Yb	U	Th
	I_{geo} value (I_{geo} Class)										
Area A	-1.37 (0)	-1.11 (0)	-1.11 (0)	-1.79 (0)	-1.64 (0)	-1.16 (0)	-1.37 (0)	-1.55 (0)	-0.75 (0)	-0.46 (0)	0.90 (1)
Area B	-1.44 (0)	-1.61 (0)	-1.28 (0)	-2.64 (0)	-2.51 (0)	-2.00 (0)	-2.10 (0)	-1.49 (0)	-0.78 (0)	-0.20 (0)	1.24 (2)
Area C	-2.20 (0)	-1.99 (0)	-2.08 (0)	-2.48 (0)	-2.41 (0)	-1.93 (0)	-1.30 (0)	-2.00 (0)	-1.19 (0)	-0.49 (0)	0.12 (1)
Area D	-0.81 (0)	-0.44 (0)	-0.66 (0)	-1.13 (0)	-1.41 (0)	-1.11 (0)	-0.47 (0)	-1.84 (0)	-0.94 (0)	0.03 (1)	0.68 (1)
Area E	-1.71 (0)	-0.71 (0)	-1.23 (0)	-1.92 (0)	-1.80 (0)	-1.49 (0)	-1.10 (0)	-1.91 (0)	-1.12 (0)	-0.23 (0)	0.41 (1)
Area F	-0.96 (0)	-0.63 (0)	-0.40 (0)	-1.25 (0)	-1.52 (0)	-1.19 (0)	-0.80 (0)	-1.93 (0)	-0.88 (0)	0.12 (1)	0.82 (1)
Area G	-0.14 (0)	-1.51 (0)	-1.51 (0)	-2.37 (0)	-1.92 (0)	-1.73 (0)	-1.43 (0)	-1.97 (0)	-1.29 (0)	-0.70 (0)	0.74 (1)
Area H	-1.09 (0)	-0.60 (0)	-0.68 (0)	-1.42 (0)	-1.66 (0)	-1.46 (0)	-1.39 (0)	-2.58 (0)	-1.36 (0)	0.20 (1)	0.69 (1)
Area I	-1.48 (0)	-1.16 (0)	-1.28 (0)	-1.26 (0)	-1.65 (0)	-5.91 (0)	-1.11 (0)	-2.10 (0)	-0.71 (0)	0.58 (1)	0.62 (1)
Area J	-2.01 (0)	-1.69 (0)	-1.78 (0)	-2.08 (0)	-2.26 (0)	-1.79 (0)	-1.54 (0)	-2.12 (0)	-1.19 (0)	-5.49 (0)	0.64 (1)
Area K	-1.97 (0)	-1.58 (0)	-1.98 (0)	-1.92 (0)	-1.98 (0)	-1.47 (0)	-1.60 (0)	-2.20 (0)	-1.17 (0)	-5.49 (0)	0.58 (1)
Area L	-1.95 (0)	-1.68 (0)	-1.68 (0)	-1.64 (0)	-1.57 (0)	-1.05 (0)	-1.07 (0)	-1.44 (0)	-0.53 (0)	0.18 (1)	1.06 (2)
Area M	-1.40 (0)	-1.00 (0)	-0.77 (0)	-1.45 (0)	-1.70 (0)	-1.55 (0)	-1.48 (0)	-2.53 (0)	-1.40 (0)	-0.47 (0)	-0.02 (0)
Area N	-0.14 (0)	0.13 (1)	-0.17 (0)	-0.46 (0)	-0.55 (0)	-0.42 (0)	-0.34 (0)	-1.04 (0)	-0.34 (0)	0.50 (1)	1.15 (2)
Area O	-1.04 (0)	-0.79 (0)	-1.04 (0)	-1.38 (0)	-1.42 (0)	-1.55 (0)	-1.20 (0)	-1.74 (0)	-1.12 (0)	-0.64 (0)	0.49 (1)
Area P	-0.53 (0)	-0.83 (0)	-0.47 (0)	-1.22 (0)	-1.50 (0)	-6.11 (0)	-1.82 (0)	-2.31 (0)	-0.69 (0)	-0.20 (0)	0.39 (1)
Area Q	-0.86 (0)	-1.06 (0)	-0.96 (0)	-1.53 (0)	-1.77 (0)	-6.09 (0)	-1.71 (0)	-2.46 (0)	-0.90 (0)	-0.50 (0)	0.18 (1)
Area R	-1.31 (0)	-0.95 (0)	-1.65 (0)	-1.87 (0)	-1.85 (0)	-1.39 (0)	-1.01 (0)	-2.42 (0)	-1.09 (0)	-0.59 (0)	0.67 (1)

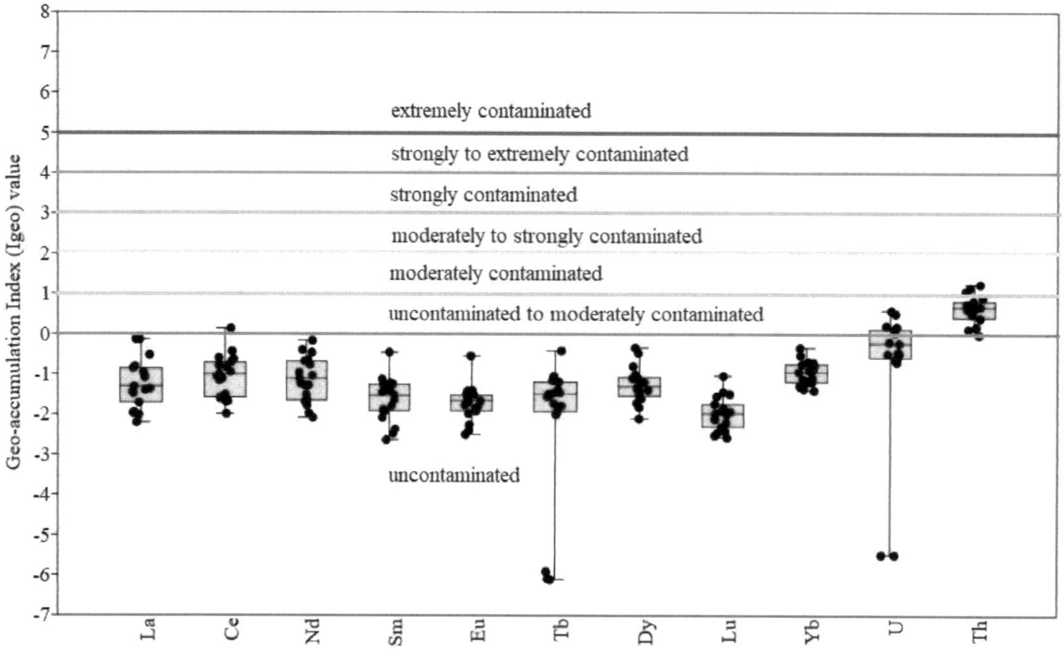

Figure 2. Geo-accumulation Index (I_{geo}) status, jitter and whisker plot of REEs, U, and Th in the soil of the Kapar industrial area

3.3 Contour map for distribution of REEs concentration in soil of Kapar industrial area

The contour maps of the distribution of REEs, U, and Th concentrations in the soil of the Kapar industrial area were depicted in Figure 3.2 (a) – (k). The groups of REEs (La, Ce, Nd, Sm, Eu, and Dy) and (Lu and Yb) were shown a similar trend of contour map. All rare earth elements showed higher concentrations in the N area compared to other areas. The N area which located in the activities of the manufacturing, food, wood, furniture, plastic, and coating industries. The distribution of U concentration showed equally distributed in all areas, within the range of 3.0 – 6.8 mg/kg except for the area of J and K slightly low concentration (Figure 3.2 (j)). The concentration distribution of Th showed higher in B area and close to mountain area (terrestrial area). The B Area was located electrical industry, cable manufacturing, aluminium smelting industry, glove, and steel industries. However, the indication from the contour map (Figure 3.2 (k)) also showed the possible sources of Th contamination most likely originated from the geogenic processes of the terrestrial such as land erosion, landslide, and land clearing for residential development.

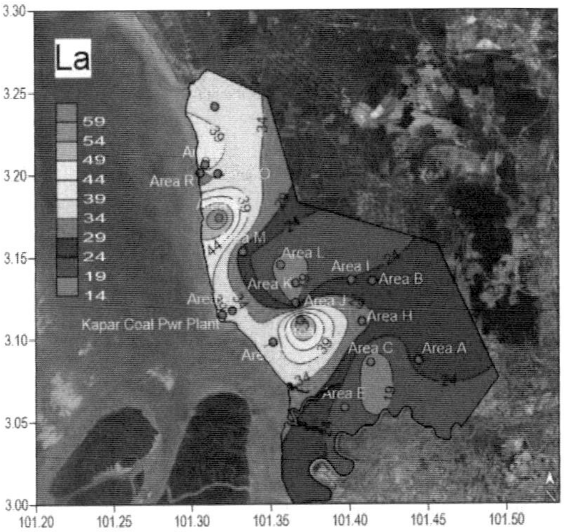
La
59
54
49
44
39
34
29
24
19
14
Kapar Coal Pwr Plant

(a)

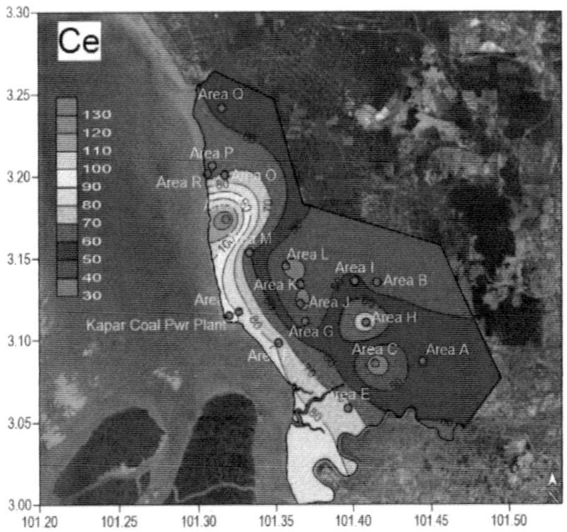
Ce
130
120
110
100
90
80
70
60
50
40
30
Kapar Coal Pwr Plant

(b)

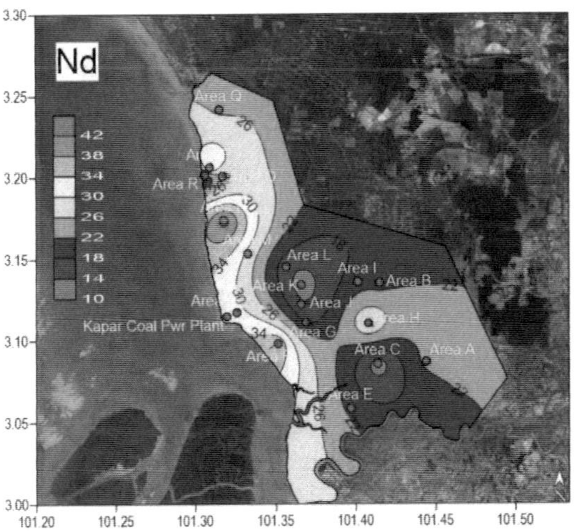
Nd
42
38
34
30
26
22
18
14
10
Kapar Coal Pwr Plant

(c)

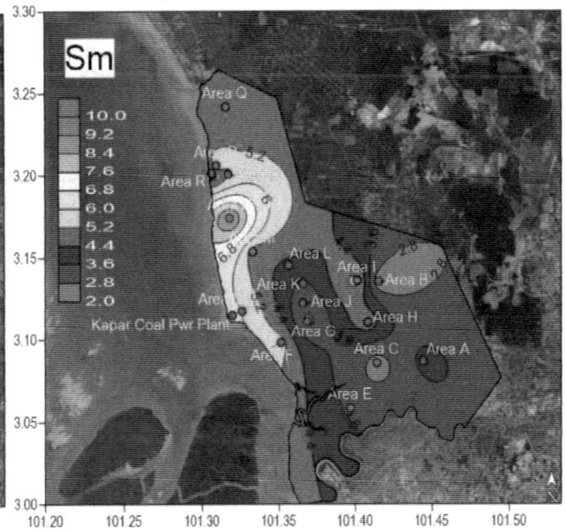
Sm
10.0
9.2
8.4
7.6
6.8
6.0
5.2
4.4
3.6
2.8
2.0
Kapar Coal Pwr Plant

(d)

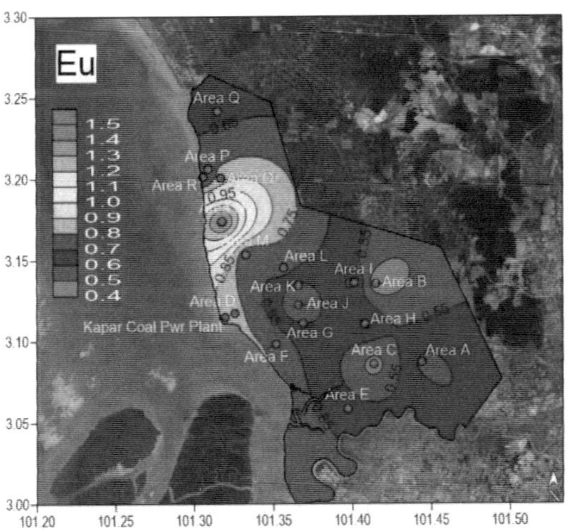
Eu
1.5
1.4
1.3
1.2
1.1
1.0
0.9
0.8
0.7
0.6
0.5
0.4
Kapar Coal Pwr Plant

(e)

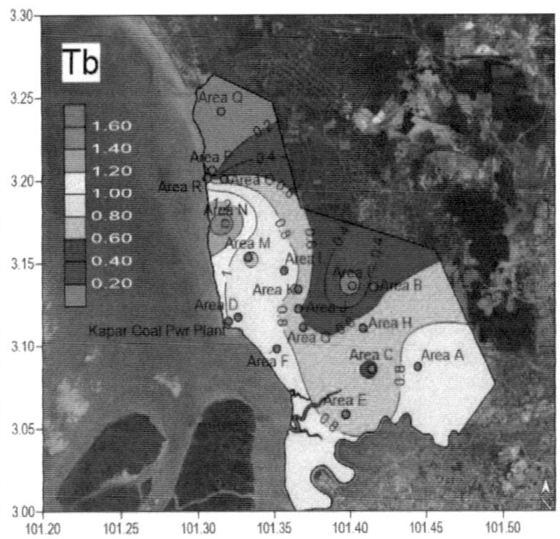
Tb
1.60
1.40
1.20
1.00
0.80
0.60
0.40
0.20
Kapar Coal Pwr Plant

(f)

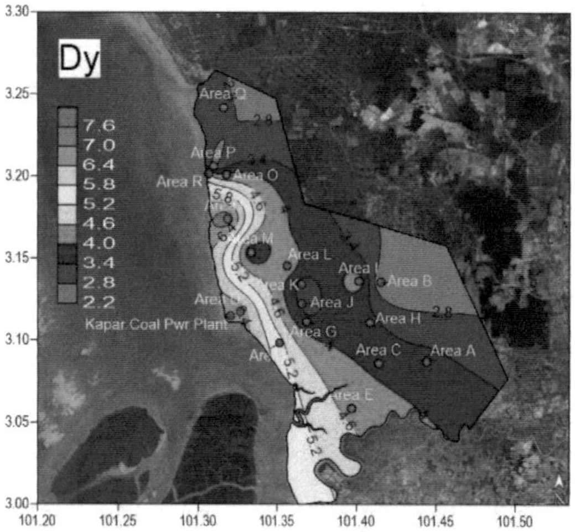

(g)

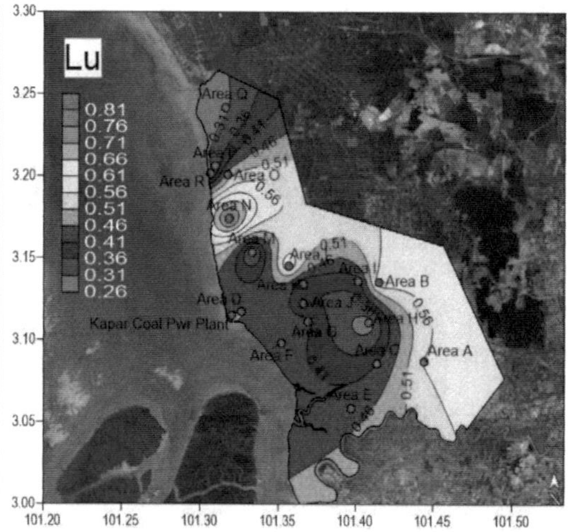

(h)

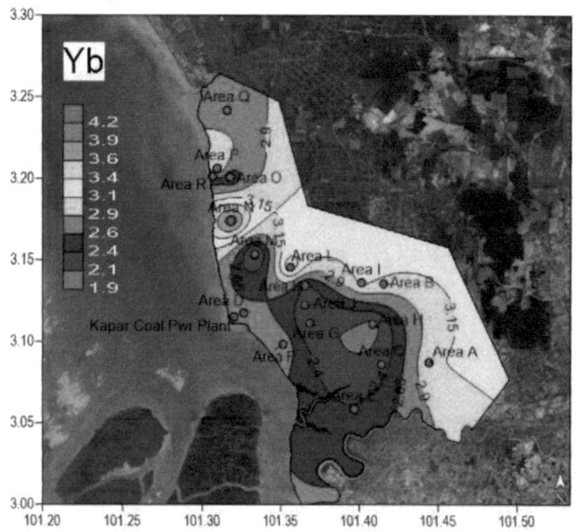

(i)

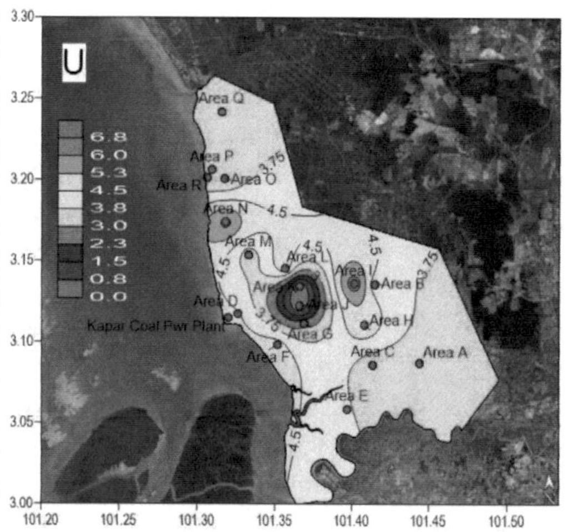

(j)

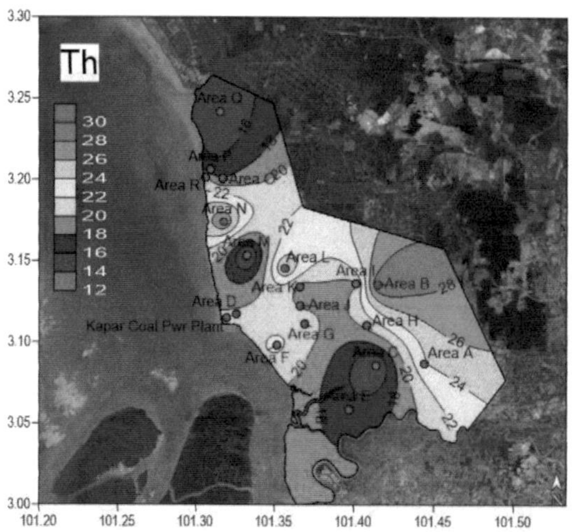

(k)

Figure 2. (a) – (k): Contour map of REEs, U and Th concentration distribution in the soil of Kapar industrial area.

4. Conclusion

The REEs such as Eu, La, Sm, Tb, Dy, Ce, Nd, Lu, and Yb showed higher in Area N compared to other areas, which is located food industry, manufacturing, wood, furniture, plastic, and coating industries. I_{geo} index values of REES in the soil of Kapar industrial were less than 0, indicating the soil can be categorised as uncontaminated with REEs. The I_{geo} index of U can be categorised uncontaminated and as uncontaminated to moderately contaminated, whilst Th can be categorised as uncontaminated to moderately contaminated. Th concentration showed higher in Area B which is located of industrial activities such as electrical, cable industry, aluminium smelter, glove, and steel industry. Besides that, other possible sources of Th could also be originated from geogenic processes such as land erosion, landslide, and terrestrial runoff.

5. References

[1] Elias M S, Wo Y M, Hamzah M S, Shukor S A, Rahman S A, Salim N A A, Azman M A and Hashim A 2017 Analysis of soil samples from gebeng area using NAA technique *AIP Conf. Proc.***1799** 030008-1-030008–11.

[2] Azman M A, Hamzah S, Rahman S A, Elias M S, Abdullah N, Hashim A, Shukor S A and Kamaruddin A H C 2015 Trace element analysis of soil type collected from the manjung and central Perak *AIP Conf. Proc.* **1659** 050003–1 – 050003–16.

[3] Sulaiman I, Omar M and Elias M . 2015 Natural Radioactivity in Soil of Sabah and Sarawak *J. Nucelar Realated Technol.* **12** 38–45.

[4] Rezaee K, Saion E B, Wood A K and Abdi M R 2010 Rare earth elements determination and distribution patterns in surface marine sediments of the South China Sea by INAA *J. Radioanal. Nucl. Chem.* **283** 823–9.

[5] Elias M S, Ibrahim S, Samuding K, Kantasamy N, Rahman S A and Hashim A 2019 Rare earth elements (REEs) as pollution indicator in sediment of Linggi River, Malaysia *Appl. Radiat. Isot.* **151.**

[6] Elias M S, Ibrahim S, Samuding K, Kantasamy N, Daung J A D, Rahman S A and Hashim A 2019 Dataset on concentration and enrichment factor of rare earth elements (REEs) in sediments of Linggi River, Malaysia *Data Br.* **25.**

[7] Alnour I A, Wagiran H, Ibrahim N, Hamzah S, Wee B S and Elias M S 2015 Rare earth elements determination and distribution patterns in granite rock samples by using INAA absolute method *J. Radioanal. Nucl. Chem.* **303** 1999–2009.

[8] Joel E S, Maxwell O, Adewoyin O O, Ehi-Eromosele C O, Embong Z and Oyawoye F 2018 Assessment of natural radioactivity in various commercial tiles used for building purposes in Nigeria *MethodsX* **5** 8–19.

[9] Elias M S, Ibrahim S, Samuding K, Rahman A R and Hashim A 2018 The sources and ecological risk assessment of elemental pollution in sediment of Linggi estuary , Malaysia *Mar. Pollut. Bull.* **137** 646–55.

[10] Muller G 1969 Index of geoaccumulation in sediments of the Rhine river *GeoJournal* **2** 108–18

[11] Abrahim G M S and Parker R J 2008 Assessment of heavy metal enrichment factors and the degree of contamination in marine sediments from Tamaki Estuary, Auckland, New Zealand *Environ. Monit. Assess.* **136** 227–38.

[12] Nobi E P, Dilipan E, Thangaradjou T, Sivakumar K and Kannan L 2010 Geochemical and geo-statistical assessment of heavy metal concentration in the sediments of different coastal ecosystems of Andaman Islands, India *Estuar. Coast. Shelf Sci.* **87** 253–64.

[13] Elias, Hamzah M S, Rahman S A, Siong W B and Salim N A A 2012 Assessment of Sediment Quality Collected from Tunku Abdul Rahman Park, Sabah *J. Sains Nukl. Malaysia* **24** 59–70.

[14] Yin H, Gao Y and Fan C 2011 Distribution, sources and ecological risk assessment of heavy metals in surface sediments from Lake Taihu, China *Environ. Res. Lett.* **6** 1–11.

[15] Zhu L, Xu J, Wang F and Lee B 2011 An assessment of selected heavy metal contamination in the surface sediments from the South China Sea before 1998 *J. Geochemical Explor.* **108** 1–14.
[16] Wedepohl K H 1995 The composition of the continental crust *Geochim. Cosmochim. Acta* **59** 1217–32.
[17] Kamber B S, Greig A and Collerson K D 2005 A new estimate for the composition of weathered young upper continental crust from alluvial sediments, Queensland, Australia *Geochim. Cosmochim. Acta* **69** 1041–58.
[18] Turekian K K and Wedepohl K H 1961 Distribution of the elements in some major units of the Earth's crust *Geol. Soc. Am. Bull* **72** 175–92.
[19] Taylor S R 1964 Abundance of chemical elements in the continental crust: a new table *Geochim. Cosmochim. Acta* **28** 1273–85.
[20] Elias M S 2021 Distribution and Source Identification of Heavy Metals, Trace and Rare Earth Elements in Surface Sediment of Linggi River *PhD Thesis* 1–309.
[21] Tavakoly Sany S B, Salleh A, Rezayi M, Saadati N, Narimany L and Tehrani G M 2013 Distribution and contamination of heavy metal in the coastal sediments of Port Klang, Selangor, Malaysia *Water. Air. Soil Pollut.* **224** 1–18.

Acknowledgement

The authors would like to thank the Malaysian Nuclear Agency and the Regional Cooperative Agreement Regional Office (RCARO) for financial support under the research grant (RCARP1/RC03). We would also like to express our gratitude and special thanks to the Analytical Chemistry Application Group staff for their sampling and analysis of samples.

IOP Conf. Series: Materials Science and Engineering 1231 (2022) 012018 doi:10.1088/1757-899X/1231/1/012018

Small beam dosimetry by using Al_2O_3 optically stimulated luminescent (OSL) dosimeters at high energy photons and electrons

Mohd Fahmi Mohd Yusof[1,(a)], Nur Amirah Shariff[1], Nur Sajidah Muhammad Fauzi[1], Ahmad Bazlie Abdul Kadir[2], Noriza Mohd Isa[2]

[1]School of Health Sciences, Universiti Sains Malaysia, 16150 Kelantan, Malaysia
[2]Health Physics and Secondary Standard Dosimetry Laboratory, Malaysian Nuclear Agency, 43000 Bangi, Selangor, Malaysia

[a]mfahmi@usm.my

Abstract. The study focused on the measurements of depth dose for small beam in the high energy photons and electrons. The percentage depth dose (PDD) was measured in 6 and 10 MV photons and 9 and 12 MeV electrons at 3 × 3 cm field sizes by using the Al_2O_3 optically stimulated luminescent (OSL) dosimeters. The PDD in OSL dosimeters were compared to those in ionization chamber and Gafchromic EBT3 film dosimeter. The results showed that the PDD in OSL dosimeters in lower photon and electron energies were in good agreement within 4% to ionization chamber and film. The PDD in 10 MV photons however showed gradual increase of deviation up to 10% beyond the depth of maximum dose (d_{max}). The surface doses in OSL dosimeters were significantly higher compared to those in ionization chamber and film dosimeter. The overall results suggested the suitability of OSL dosimeters to be used as indirect dosimetry works in high energy photons and electrons.

1. Introduction

The use of small fields in radiotherapy helped in providing better accuracy of therapeutic beam deliveries. They usually having dimensions smaller than the lateral range of the charged particles that contribute to the dose deposited at a point along the central. The field sizes of less than 3 cm × 3 cm are considered to be small for photons beam [1]. It can be obtained by using multi leaf collimator (MLC) in typical linear accelerators. The MLC enable to conform with shape of target tumour and tailor to the prescribed dose distribution to the planning target volume (PTV). Not all dose distribution inside PTV or another volume are homogenous but heterogonous dose distribution are commonly occurred [2]. Therefore, the smaller and irregular collimator are required for complex dose distribution to ensure that the planned dose distributions are accurate throughout the define tumour volume. The small beam in radiotherapy is commonly applied in various treatment methods such as intensity-modulated radiotherapy (IMRT), volumetric-modulated arc therapy (VMAT), stereotactic radiotherapy (SRT), stereotactic radiosurgery (SRS) and stereotactic ablative radiotherapy (SABR) [3]. The challenges in small field dosimetry includes lateral electronic disequilibrium, source occlusion, volume averaging effect and beam alignment [4-6]. Several dosimeters such as MOSFET, TLDs, ionization chambers as well as dose simulations by the treatment planning software (TPS) commonly used in radiotherapy.

The use of Al_2O_3 optically stimulated (OSL) dosimeters for dosimetry works in medical physics has been increasing in the recent years [7-9]. Previous works also indicated the suitability of OSL dosimeters for dosimetry works in the open fields of high energy photons and electrons [10,11].
The measurement of percentage depth dose (PDD) had been the most convenient method to evaluate the attenuation and dosimetric properties of dosimeters at the function of depth in clinical photons and electrons [12]. This study measured the PDD of small beams of photons and electrons by using the OSL dosimeter in comparison to the common methods of ionization chamber and radiochromic film dosimeter.

2. Methodology

2.1. Experimental set up

The OSL dosimeters were irradiated at high energy photon of 6 and 10 MeV and electrons of 9 and 12 MeV by using Siemen PRIMUS linear accelerator in Nuclear Medicine, Radiotherapy and Oncology Department, Hospital University Sains Malaysia (HUSM) as shown in Figure 1. The solid water phantoms were used to simulate a human soft tissues. A number of three OSL dosimeters were placed at the central axis of the beam at 3 × 3 cm field size to simulate the small beams. The small beam in electrons were achieved by using an applicator of 3 × 3 cm. A bolus was always placed on top of the OSL dosimeters at depths to minimize the air gaps between the OSL dosimeters and solid water phanom slabs. The exposures were made at 100 cm source to surface distance (SSD) and 100 cGy dose indicated by the monitor unit (MU). The depth doses were measured from the surface until approximately 15 cm and 5 cm depths for high energy photons and electrons respectively while maintaining the SSD.

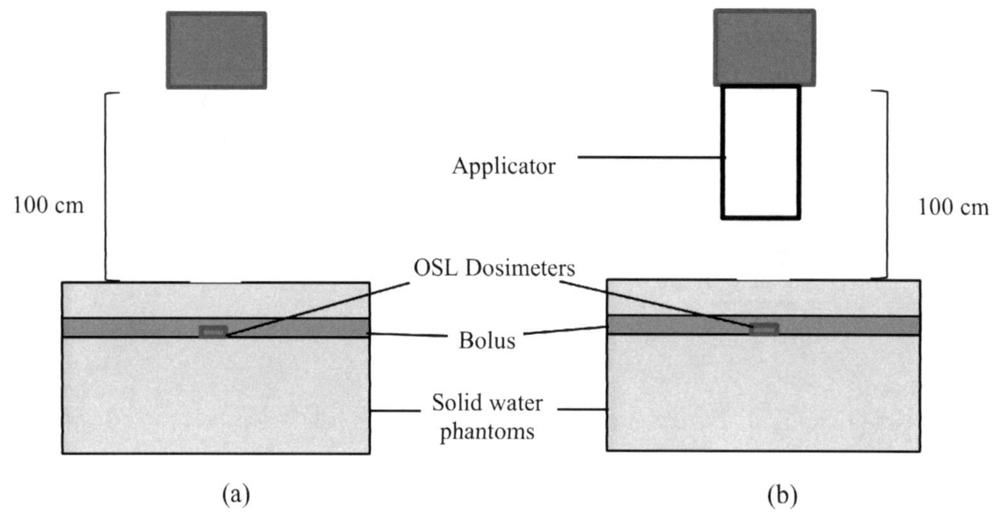

Figure 1. The experimental set up for the irradiation of OSL dosimeters at (a) high energy photons and (b) electrons.

2.2. Data Analysis

The signals of the OSL dosimeters were analysed by using OSL reader model MicroStar InLight at the Malaysian Nuclear Agency (Nuclear Malaysia) [13]. The OSL signals were obtained by calculating the ratio between the photomultiplier tube (PMT) count with the calibration factor given by the equation:

$$Raw\ Dose\ (mGy) = \frac{PMT\ Count}{Calibration\ Factor\ x\ Sensitivity} \quad (1)$$

The dose reading in the OSL dosimeter is obtained by calculating the difference between the initial signal before irradiation with the signal after the irradiation given by the equation

$$Dose_{OSL} = (Raw\ Dose\ - Initial\ Dose) - Control\ Dose \tag{2}$$

with *Control Dose* is the reading in the unexposed OSL dosimeter of the same group.

The percentage depth dose in all dosimeters were calculated by using the equation:

$$PDD = \frac{Dose\ at\ any\ depth, d}{Dose\ at\ maximum\ depth, D_{max}} x\ 100\% \tag{3}$$

The *PDD* at small beams in high energy photons and electrons were also measured by using ionization chamber and Gafchromic EBT3 radiochromic film dosimeters. The *PDD* curves by the OSL dosimeters were plotted and compared to the *PDD* in ionization chamber and EBT3 film at all experimented energies of photons and electrons.

3. Results and discussion

The plotted *PDD* curves of OSL dosimeters at small beam of high energy photons in comparison to the ionization chamber and EBT3 film dosimeters are shown in Figure 2. The *PDD* curves of OSL dosimeters at 6 MeV photons were in excellent agreement to those in ionization chamber and EBT3 film as shown in Figure 2(a) with maximum percentage of discrepancies of 4.22 and 1.31% in comparison to the ionization chamber and EBT3 film respectively. The results were in good agreement with the previous works that measured the *PDD* of OSL in high energy photons of open field of 10 × 10 cm [10,11]. The *PDD* in 10 MeV photons however showed higher percentage of difference between all dosimeters. The *PDD* in OSL dosimeters showed maximum percentage of discrepancies of 27.6 and 57.4% in comparison to ionisation chamber and EBT3 film respectively. The results were also consistent with the previous works that indicated higher percentage of discrepancies of OSL readings to those in ionisation chamber and film dosimeters at high energy photons of open field of 10 × 10 cm 10,11]. Higher percentage of differences were observed in higher energy photons due to higher production of ion pairs within the medium that would have different detection among the dosimeters. The percentage difference were more significant between OSL and TLD at higher photon energies due to higher probability of signal fading in the TLD compared to that in OSL [14].

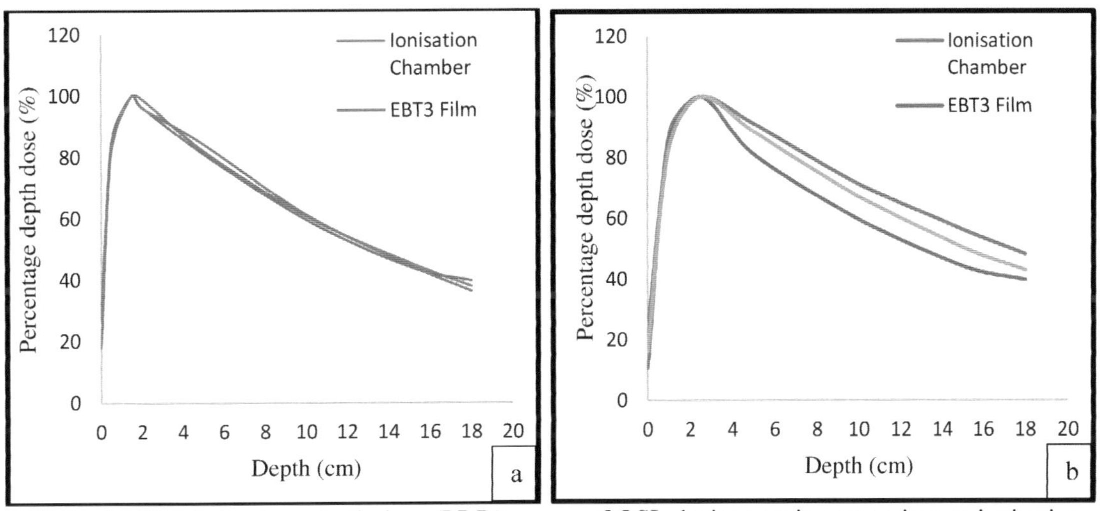

Figure 2 The percentage depth dose (PDD) curves of OSL dosimeters in comparison to ionisation chamber and EBT3 film in (a) 6 MeV and (b) 10 MeV photons

The PDD curves of OSL in electrons in comparison to ionisation chamber and EBT3 film dosimeters are illustrated in Figure 3. The PDD in OSL showed maximum percentage of discrepancies of 10.5 and 13% when measured at 6 MeV electrons in comparison to ionisation chamber and EBT3 film respectively as shown in Figure 3(a). This is due to the high dispersion of electrons at lower energies that would affect the shape of the PDD curves [12]. The PDD in OSL at 12 MeV electrons on the other hand showed lower percentage of discrepancies of below 5% in comparison to ionisation chamber and EBT3 films as shown in Figure 3(b). This is due to the less dispersion of electrons at higher electron energies [12]. The surface dose in OSL showed an excellent agreement within 3% variation in comparison to ionisation chamber and EBT3 films at all experimented electron energies.

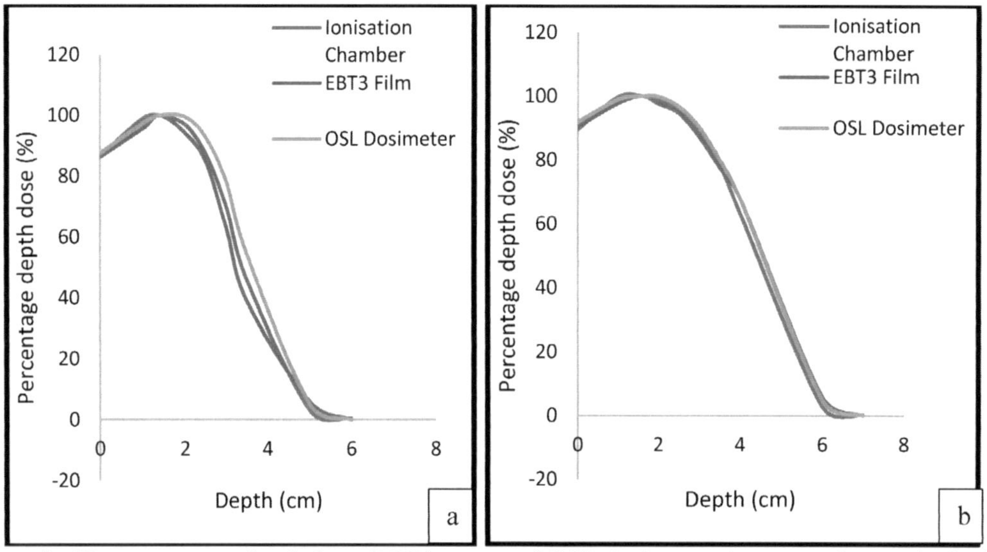

Figure 3 The percentage depth dose (PDD) curves of OSL dosimeters in comparison to ionisation chamber and EBT3 film in (a) 9 MeV and (b) 12 MeV electrons

The electron beam parameters of electrons in OSL dosimeters, ionization chamber and EBT3 films is shown in Table 1. The electron parameters of OSL dosimeters showed closer values to those in ionization chamber and film dosimeters when measured in higher electron energies. The results very much consistent to the PDD curves of the OSL dosimeters on which the PDD curves in higher electron energies showed better agreement and lower percentage of variations. The electron parameters are often used in radiotherapy treatment planning involving electrons on which the therapeutic range (R_{80} and R_{90}) is used to determine the maximum dose to the tumor volumes.

Table 1. The electron beam parameters in OSL dosimeters in comparison to ionisation chamber and EBT2 film dosimeters in electrons of 9 and 12 MeV.

Electron parameters	**Depth (cm)** **OSL Dosimeters**	**Ionisation chamber**	**EBT3 films**
	9 MeV		
R_{90}	2.59	2.35	2.21
R_{80}	2.94	2.72	2.60
R_{50}	3.63	3.42	3.27
	12 MeV		
R_{90}	3.02	2.93	2.82
R_{80}	3.51	3.43	3.38
R_{50}	4.58	4.44	4.59

A paired sample t-test was conducted between OSL dosimeters and ionization chamber and EBT3 films in all experimented energies of photons and electrons as shown in Table 2 and 3. The results of the paired sample t-test were consistent with the PDD curves as illustrated in Figure 2 and 3. In photons, the depth doses in OSL had no significant differences in comparison to ionisation chamber and EBT3 films when measured at lower photon energies indicated by the *p*-values (Sig. 2-tailed). The depth doses at higher energy photons however showed significant differences in comparison to ionisation chamber and EBT3 films. The depth dose in electrons however shoed no significant differences between OSL dosimeters with ionisation chamber and EBT3 films when measured at higher electron energies. The depth dose at lower electron energies however showed significant differences between OSL dosimeters with ionisation chamber and EBT3 films.

Table 2. The paired sample t-test between the depth doses in OSL dosimeters and ionisation chamber at 6 and 10 MeV photons and 9 and 12 MeV electrons.

Energy	Paired Differences					t	df	Sig. (2-tailed)
	Mean	Std. Deviation	Std. Error Mean	95% Confidence Interval of Differences				
				Lower	Upper			
6 MV	0.9187	4.12767	1.45935	-2.53207	4.36957	0.630	7	0.549
10 MV	-2.38571	4.02189	1.52013	-6.10535	1.33392	-1.569	6	0.168
9 MeV	2.68556	3.74496	1.24832	11.9308	5.56419	2.151	8	0.065
12 MeV	5.3811	6.24147	2.02049	5.8349	10.17873	2.586	8	0.532

Table 3. The paired sample t-test between the depth doses in OSL dosimeters and EBT3 film dosimeters at 6 and 10 MeV photons and 9 and 12 MeV electrons.

Energy	Paired Differences					t	df	Sig. (2-tailed)
	Mean	Std. Deviation	Std. Error Mean	95% Confidence Interval of Differences				
				Lower	Upper			
6 MV	0.67238	3.07102	1.08577	-1.89507	3.23982	0.619	7	0.555
10 MV	-2.96000	7.59024	2.86884	-9.97980	4.05980	-1.032	6	0.342
9 MeV	1.32000	1.59030	0.50290	0.18237	2.45763	2.625	9	0.028
12 MeV	1.17500	1.31250	0.41505	0.23610	2.11390	2.831	9	0.520

4. Conclusion

The OSL dosimeters showed good depth dose measurements in small beams of high energy photons and electrons. The depth dose of OSL dosimeters showed closer depth dose values to ionisation chamber and EBT3 film dosimeters when measured in lower photon energies and higher electron energies. The electron beam parameters in OSL dosimeters also showed good agreement to those in ionisation chamber and EBT3 films. The overall results indicated the suitability of OSL dosimeters to be used and measurement method for small beams in radiotherapy in the applications of the stereotactic radiotherapy and radiosurgery.

5. References

[1] Charles P H, Cranmer-Sargison G, Thwaites D I, Crowe S B, Kairn T, Knight R T and Trapp J V 2014 A practical and theoretical definition of very small field size for radiotherapy output factor measurement *Med. Phys.* **41(4)** p 041707.

[2] Bagheri H, Soleimani A, Gharehaghaji N, Mesbahi A, Manouchehri F, Shekarchi B, Dormanesh B and Dadgar H A 2017 An overview on small-field dosimetry in photon beam radiotherapy: Developments and challenges *J. Canc.Res. Therapet.* **13 (2)** p 175-185.

[3] Palmans H, Andreo P, Huq M S, Seuntjens J, Christaki K E and Meghzifene A 2018 Dosimetry of small static fields used in external photon beam radiotherapy: Summary of TRS-483, the IAEA–AAPM international Code of Practice for reference and relative dose determination. *Med. Phys.* **45(11)** p 1123-1145.

[4] Paskalev K A, Seuntjens J P, Patrocinio H J and Podgorsak E B 2003 Physical aspects of dynamic stereotactic radiosurgery with very small photon beams (1.5 and 3 mm in diameter). *Med. Phys.* **30(2)** p 111–118.

[5] Tsiakalos M F, Theodorou K, Kappas C, Zefkili S, and Rosenwold J C 2004 Analysis of the penumbra enlargement in lung versus the quality index of photon beams: a methodology to check the dose calculation algorithm. *Med. Phys.* **31(4)** p 943-949.

[6] Mesbahi A 2008 The effect of electronic disequilibrium on the received dose by lung in small fields with photon beams: Measurements and Monte Carlo study. *Iran. J. Radiat. Res.* **6(2)** p 70-76.

[7] Kobayashi I, Sekiguchi H, Tatsuta F, Komori H, Koga T and Morishima H 2005 The environment dose measurement using the OSL dosimeter. *Int.l Congr.* **1276** p 305-306.

[8] Tien C J, Ebeling R, Hiatt J R, Curran B, and Sternick E 2012 Optically Stimulated Luminescent Dosimetry for High Dose Rate Brachytherapy. *Front. Oncol.* **2(91)** p 1-7.

[9] Knežević Ž, Stolarczyk L, Bessières I, Bordy J, Miljanić S and Olko P 2013 Photon dosimetry methods outside the target volume in radiation therapy: Optically stimulated luminescence (OSL), thermoluminescence (TL) and radiophotoluminescence (RPL) dosimetry. *Radiat. Meas.* **57** p 9-18.

[10] Yusof M F M, Joohari N A, Abdullah R, Shukor N S A, Kadir A B A and Isa N M 2018. Dosimetry of Al_2O_3 optically stimulated luminescence dosimeter at high energy photons and electrons. *IOP Conf. Ser.: Mater. Sci. Eng.* **298** p 012023.

[11] Yusof M F M, Othman N N M, Abdullah R, Kadir A B A, Salleh H and Isa N M 2020 Percentage depth dose measurement in high energy photons and electrons by using the Al_2O_3 optically stimulated luminescent (OSL) dosimeter *IOP Conf. Ser.: Mater. Sci. Eng.* **785** p 012002.

[12] Khan M F 2010 The physics of radiation therapy (Philadelphia: Lippincott William &Wilkins)

[13] Landauer 2019 MicroSTAR®: the mobile radiation dosimeter reader.

[14] Vasilache R A, Nicola S T and Tiron R 1999 Characteristics of TLD-100 fading and its influence on the calibration of personal dosemeters. *Radiat. Prot. Dosim.* **85(1)** p 183-185.

Acknowledgement

The authors would like to acknowledge the financial support from USM Research Universiti (RU-Individual) Grant no. 1001/PPSK/8012381. The authors also acknowledge the Nuclear Medicine, Radiotherapy and Oncology Department, Hospital Universiti Sains Malaysia for the use of linear accelerator in this research.

IOP Conf. Series: Materials Science and Engineering 1231 (2022) 012019 doi:10.1088/1757-899X/1231/1/012019

Applying empirical method for performance metrics measurement of a nuclear security system

Maizura Ibrahim[1,a)], Siti Nurbahyah Hamdan[1], Saa'idi Ismail[1], Mohamad Safuan Sulaiman[1], Sufian Norazam Mohamed Aris[1], Mohd Hasnor Bin Hasan[1], Mohd Dzul Aiman Aslan[1], Nur Fatini Abdul Ghani [1] and Amy Hamijah Ab Hamid[1]

[1]Information Technology & Application Centre, Technical Support Division, Malaysian Nuclear Agency, Bangi, 43000 Kajang, Selangor Darul Ehsan, Malaysia.

[a]maizura@nm.gov.my

Abstract. A nuclear security system is often developed based on a particular framework. In our case, a nuclear security system is developed based on the big data analytics framework with the purpose of integrating all nuclear security systems within a nuclear facility. Nuclear security can be defined as the prevention and detection of, and response to, theft, sabotage, unauthorized access, illegal transfer, or other malicious acts involving nuclear material, other radioactive substances, or their associated facilities. Whereas, big data analytics is denoted as a new way of collecting, analyzing large amounts of data, finding the appropriate patterns to support decision making, hence improving the action taken in solving the problems. The system-based evaluations methodology is widely used to measure the performance metrics of nuclear security systems. However, the system-based evaluation methodology consumes time and demands a lot of effort because in applying the methodology, a complete nuclear security system that is based on the particular framework needs to be designed, developed, and implemented in the nuclear facilities environment before the performance metrics could be measured. In this paper, we present an empirical method of measuring the performance metrics by utilizing the Rasch Measurement Model. The methodology is formulated to reduce the time and efforts in measuring the performance metrics of a nuclear system that is developed based on a framework...

1. Introduction

Nuclear security is defined as the prevention and detection of, and response to, theft, sabotage, unauthorized access, illegal transfer, or other malicious acts involving nuclear material, other radioactive substances, or their associated facilities[1]. Any system that is created and implemented to achieve nuclear security, whether it is a computer-based system or a manual system is regarded as a nuclear security system. The context of our work is derived based on a computer-based system, which in broader terms is also known as the electronic management system. Thus, the nuclear security system focused in our work is regarded as the application/software that is developed to provide the prevention and detection of, and response to, theft, sabotage, unauthorized access, illegal transfer, or other malicious acts involving nuclear material, other radioactive substances, or their associated facilities. Our work uses the big data analytics framework as a basis for system design[2]. We defined Big data as the high volume, high velocity, high variety, and high veracity, i.e., the 4Vs of data (information assets) that require new forms of processing, to enable enhanced decision making, insight discovery, and process optimization [3].

Several methods have been used to evaluate the performance metrics of a system, such as the system-based, theoretical-based, and hypothetical-based methodology. The system-based methodology is widely used in measuring the performance metrics of an application/software in the Information Technology field. Whereas, the theoretical-based and hypothetical-based methodologies are widely applied in measuring the performance metrics of a system in the Nuclear Security field. The system-based evaluation methodology is defined as performing several systematic processes into the final version of a system as an assessment to see whether the system has met the specified goals or not. The system-based evaluation methodology consumes time and demands a lot of effort. Therefore, in this paper, we present an empirical method of measuring the performance metrics by utilizing the Rasch Measurement Model. The methodology is formulated to reduce the time, efforts, and process complexity in measuring the performance metrics of a nuclear system that is developed based on a framework. The first section has introduced the topic addressed in this paper. The second section reveals the related works. The third section presents the proposed methodology. The section is followed by a comparison and discussion of three different methodologies that are commonly applied. The final section presents the conclusions of the paper and proposes future work.

2. Related works

This paper reports on a review of methodologies that are applied in the assessment process of a nuclear security system. In addition, this article also presents the technique of applying an empirical method in measuring the performance metrics of our work that were previously published in [2][4][5][6][7]. While the system-based methodology is commonly used to evaluate the application/software in the IT domain, the theoretical-based and hypothetical-based techniques are commonly used in the Nuclear Security domain. However, for this work, we proceed the research through the path that is commonly applied in software engineering research. In other words, we look at our research work from the perspective of an empirical research project. Therefore, for the evaluation process, the empirical method is proposed to be applied in assessing our work. The empirical method is defined as any procedure for conducting an investigation that relies upon experimentation and systematic observation rather than theoretical speculation.

The system-based methodology is widely applied in assessing an application software following a regular application development process such as software development lifecycle[8]. In system-based evaluation methodology, the application software should be designed and developed first before the assessment process could be conducted. Apart from that, the theoretical-based and hypothetical-based are commonly found in the literature related to nuclear security assessment methodologies. The nuclear security assessment methodologies and tools are designed to provide assurance that regulated facility protection measures are effective against the threat as defined in the State's threat assessment or design basis threat. The IAEA-TECDOCS-1868 entitled the Nuclear Security Assessment Methodologies for Regulated Facilities (NUSAM) applied the theoretical-based and hypothetical-based methodology [9]. This IAEA-TECDOCS was published to provide the Member States of IAEA with a standard nuclear security assessment methodology and criteria to assist in assessment tool(s) selection. The IAEA Nuclear Security Series No. 4 entitled Engineering Safety Aspects of the Protection of Nuclear Power Plants Against Sabotage applied the hypothetical-based methodology intending to prevent the revelation of the detailed design of the nuclear security system [10].

3. The empirical methodology for measuring the performance metrics of a nuclear security system based on big data analytics framework

The proposed empirical methodology used to evaluate the performance metrics for the big data analytics nuclear security framework is depicted in Fig. 1. The methodology outlined nine major steps. The first step is the framework that will be used as the basis for the application/software development must be developed. For the framework development methodology, we left the option open, which means any existing methods could be adopted. The second step is the process of identifying the framework's peculiarities elements with regards to the previous nuclear security framework. In our work, the fundamental elements of our framework are based on the existing International Legal Framework for Nuclear Security, which we constituted as a non-big data analytics nuclear security framework. The International Legal Nuclear Security Framework is illustrated by the International Atomic Energy Agency (IAEA) in multiple publications of the IAEA Nuclear Security Series, staggered in four different levels of grouped materials, i.e. Nuclear Security Fundamentals [11], Recommendations, Implementing Guides, and Technical Guidance, Reference Manuals, Training Materials. Therefore, the peculiarities elements are identified as the elements that do not appear in the existing International Legal Framework for Nuclear Security. These peculiar elements are the performance metrics that are going to be measured during the evaluation process.

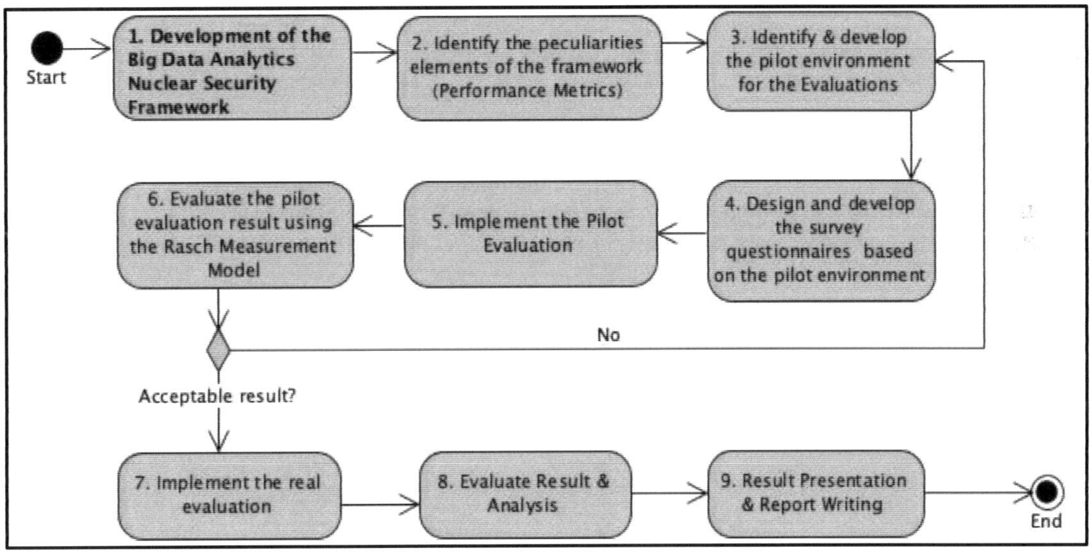

Figure 1. An empirical method for measuring the performance metrics of a nuclear security system based on a big data analytics framework.

The third step is to identify and develop the pilot environment for the evaluations processes. This is the preliminary step taken to make sure the instrument used in the measurement, evaluation, and statistical analysis before any data collection hence data analysis, is correct and could measure what the evaluation process suppose to measure. This is a very important step that needs to be given very serious consideration because a faulty instrument will render the data useless and the data analysis is therefore null and void. During this step, the research questions are also formulated and clarified as the elementary step before the design and development of the survey questionnaire (step 4) because research design is a key element in shaping the way research questions are going to be asked in the survey questionaire[12]. The broad research question we ask in our work is, to what extent does adopting our big data analytics nuclear security framework affect the security of radioactive materials in the Malaysian Nuclear Agency? From the broad research question, variables are identified and distinguished between three types of variables, which are dependent, independent, and intervening.

Step 4 consists of activities to design and develop the survey questionnaire based on the focus and clarified research questions, the dependent, independent, and intervening variables. Here we depict this research as a case study method because this research does not fundamentally rely on comparing cases but on fully understanding the 'wholeness' effect of implementing the big data nuclear security framework on securing the radioactive/nuclear materials in the Malaysian Nuclear Agency. We also consider our research questions as a descriptive type rather than an explanatory one because what we want from the result of this survey is to describe and confirm the effect of implementing the big data nuclear security framework specifically in the Malaysian Nuclear Agency, based on the cause and effect that are theoretically mentioned in previous reports published by World Institute of Nuclear Security (WINS) [13][14]. Our survey questionnaire is developed by making use of the Google form online survey platform.

After step 4 is completed, the implementation is done by distributing the online survey questionnaire to the pilot experiment targeted individuals via email and WhatsApp messaging platform (step 5). In step 6, the result is analyzed based on the procedure outlined by Rasch Measurement Model[15][16][17]. Rasch model provides a way to evaluate the instruments used during the pilot experiment. The results obtained during the pilot experiment could be used to validate the research design. If the obtained results are acceptable, then one could proceed to the execution of step 7, step 8, and step 9. If the pilot results point out that the instrument used is not valid, the procedure in step 3 should be re-evaluated, then proceed with pilot implementation again, until the results are acceptable. We have several reasons and justification for why we propose our method for the evaluation of our work. The next section compares and discusses the reasons.

4. Methodologies comparison and discussion

Table 1 presents the result of our review process that is done to compare the empirical-based evaluation method with the theoretical/hypothetical-based, and system-based. The comparison is made based on three properties, i.e., requirements, time consumption, risk, and process complexity.

Table 1. Comparison of the methodologies.

Properties	Empirical-Based	System-Based Evaluation Method	Theoretical / Hypothetical-Based
Requirements	A framework that is used as the basis for the applications/systems development must be developed first.	A complete system needs to be designed, developed, and implemented in the nuclear facilities environment before the performance metrics could be measured.	A complete design of the system. Most designs are developed based on threat assessment (DBT), thus DBT must be developed first.
Time consumption	Low	Long	Moderate
Risk	Low	High	Low
Process complexity	Low	High	High

The requirements are defined as the input that is needed to proceed with the evaluation process following each specific methodology. To start the evaluation process by following our empirical-based method, a framework that is used as the basis for the application/system development needs to be developed and ready first. Whereas, to adopt the system-based evaluation method, a complete nuclear security system needs to be designed, developed, and implemented in the nuclear facilities before the evaluation process could be started. Apart from that, to evaluate the nuclear security system following the theoretical/hypothetical-based methodology, a complete system design based on threat assessment (Design Basis Threat) needs to be ready before the evaluation process could be initiated. Therefore, requirements for implementing empirical-based methodology constitute low time consumption, low risk, and low process complexity. However, we regard the system-based evaluation method consumes a long time due to the requirement of system design, development, and implementation that commonly take a very long time to be finished. Hence, the risk of applying this method is regarded as high because if the system development fails, the chances of failing to obtain the performance evaluation result are very high. Thus, the context of risk here meant the risk of not being able to start the evaluation process due to the fail of preparing their requirements. Meanwhile, the context of process complexity in this paper is regarded as the overall process from preparing the requirements until the evaluation is finished. Since the system-based evaluation method and the theoretical/hypothetical-based method require more than one stage of preparation before the evaluation process could be initiated, they are regarded as a high complexity process compare to the empirical-based method which only needs a framework preparation (one stage only).

5. Conclusion and future works

In this paper, we had proposed an empirical methodology to evaluate the performance metrics of a computer-based nuclear security system that is developed based on the big data analytic framework. The key element of this methodology is that this methodology is developed by applying the empirical method, so there will be no probabilistic assessment involved. Besides, the implementation process is straightforward and could be done by distributing and collecting the result via the online survey questionnaire, thus, it could be re-implement on any online survey platform. The design of the survey and analysis is done by applying Rasch Measurement Measure Model, which constitutes a precise cycled evaluation. This methodology has low time consumption, low risk, and low process complexity compared to the previous methodologies such as system-based, theoretical-based and hypothetical-based methodology. Our future works are to implement the proposed methodology after the method validation process is done by the expert in our research collaboration team members.

6. References

[1] International Atomic Energy Agency 2008 Security in transport of radioactive material, implementing guide *IAEA Nuclear Security Series No.9,* (Vienna, Austria).

[2] Ibrahim M, Ismail S, Haris M F, Hamdan S N, Sulaiman M S, Hamid A H A, Aris S N M, Aslan M D A, Ghani N F A, and Hassan H 2021 Big Data analytics nuclear security framework, *IOP Conf. Ser.: Mater. Sci. Eng.* **1106** 012026

[3] Beyer M A and Laney D 2012 The importance of big data: A definition (Stamford, CT: Gartner).

[4] Ibrahim M, Ismail S, Haris M F, Hamdan S N, Sulaiman M S, Hamid A H A, Aris S N M, Aslan M D A, Ghani N F A, and Hassan H 2017 2017 A proposal on adaptation of big data analytics framework in nuclear security toward development of integrated, real-time, and intelligent nuclear security system, *Nuclear R & D Seminar, Selangor, Malaysia.*

[5] Ibrahim M, Ismail S, Haris M F, Hamdan S N, Sulaiman M S, Hamid A H A, Aris S N M, Aslan M D A, Ghani N F A and Hassan H 2018 The integrated intelligent nuclear security (I2NS) system: A progress report for 2018, *Technical Report No.NUKLEARMALAYSIA/L/2018/173* (Kajang, Selangor, Malaysia: Malaysian Nuclear Agency).

[6] Ibrahim M, Ismail S, Haris M F, Hamdan S N, Sulaiman M S, A. H .A. Hamid, S. N. M. Aris, M. D. A. Aslan, N. F. A. Ghani, and H. Hassan 2020 Malaysian Nuclear Agency integrated & intelligent nuclear security (I2NS) system: A progress report 2019 *Technical Report No.NUKLEARMALAYSIA/L/2020/10* (Kajang, Selangor, Malaysia: Malaysian Nuclear Agency).

[7] Ibrahim M, Ismail S, Haris M F, Hamdan S N, Sulaiman M S, Hamid A H A, Aris S N M, Aslan M D A, Ghani N F A and Hassan H 2020 Big data analytics for nuclear security: A literature review, *Nuclear R & D Seminar 2020 16th – 19th November 2020, Kajang, Selangor, Malaysia.*

[8] Unit Pemodenan Tadbir dan Perancangan Pengurusan Malaysia (MAMPU) 2019 KRISA: Kejuruteraan sistem aplikasi sektor awam-satu panduan komprehensif dan praktikal, (Putrajaya, Malaysia: MAMPU).

[9] International Atomic Energy Agency (IAEA) 2019 The nuclear security assessment methodologies for regulated facilities (NUSAM) *IAEA-TECDOCS-1868* (Vienna, Austria).

[10] International Atomic Energy Agency IAEA 2007 Engineering safety aspects of the protection of nuclear power plants against sabotage *IAEA Nuclear Security Series No.4* (Vienna, Austria).

[11] International Atomic Energy Agency (IAEA) 2013 Objective and essential elements of a state's nuclear security regime, nuclear security fundamentals *IAEA Nuclear Security Series No.20* (Vienna, Austria).
[12] David de Vaus 2002 Surveys in social research 5th ed (NSW, Australia: Allen & Unwin).
[13] World Institute of Nuclear Security (WINS) 2015 Data analytics for nuclear security, how real-time integrated data management could support nuclear security *WINS Special Report Series, January 2015* (Vienna, Austria).
[14] World Institute of Nuclear Security (WINS) 2015 Big data in motion, how real-time integrated data management could support nuclear security *WINS Special Report Series, September 2015*, (Vienna, Austria).
[15] Bond T G and Fox C M 2015 Applying the Rasch model: Fundamental measurement in the human sciences 3rd ed. (Routledge, NY, US: Taylor & Francis Publishers).
[16] Wright B. and Mok M 2004 An overview of the family of rasch measurement models, in introduction to Rasch measurement: theory, models, and applications ed Everett J, Smith V and Smith R M.
[17] Wright B D and Stone M H 1999 Measurement essentials 2nd ed (Wilmington, Delaware, USA: Wide Range, Inc).

Acknowledgments
This study is supported by the International Atomic Energy Agency (IAEA) research grant under the IAEA Coordinated Research Project CRP J02011: Improving the Security of Radioactive Material throughout its Lifecycle, Associated Facilities, and Associated Activities (Contract No. 22165).

AUTHOR INDEX